This book is a self-contained introduction to key topics in advanced general relativity. The opening chapter reviews the subject, with strong emphasis on the geometric structures underlying the theory. The second chapter discusses 2-component spinor theory, its usefulness for describing zero-mass fields and its practical application via Newman-Penrose formalism, together with examples and applications. There follows an account of the asymptotic theory far from a strong gravitational source, describing the mathematical theory by which measurements of the far-field and gravitational radiation emanating from a source can be used to describe the source itself. Finally, the characteristic initial value problem is described, first in general terms, and then with particular reference to relativity, concluding with its relation to Arnol'd's singularity theory. Exercises are included throughout.

CAMBRIDGE MONOGRAPHS ON
MATHEMATICAL PHYSICS

General editors: P. V. Landshoff, D. R. Nelson, D. W. Sciama, S. Weinberg

ADVANCED GENERAL RELATIVITY

Cambridge Monographs on Mathematical Physics

ADVANCED GENERAL RELATIVITY

JOHN STEWART

Department of Applied Mathematics and Theoretical Physics
University of Cambridge

CAMBRIDGE
UNIVERSITY PRESS

Published by the Press Syndicate of the University of Cambridge
The Pitt Building, Trumpington Street, Cambridge CB2 1RP
40 West 20th Street, New York, NY 10011, USA
10 Stamford Road, Oakleigh, Melbourne 3166, Australia

First published 1991
First paperback edition 1993

British Library cataloging in publication data
Stewart, John *1943–*
Advanced general relativity
1. Physics.   General theory of relativity
I. Title
530.11

ISBN 0-521-32319-3 hardback
ISBN 0-521-44946-4 paperback

Library of Congress cataloging in publication data available

Transferred to digital printing 2003

CONTENTS

PREFACE

General relativity is the flagship of applied mathematics. Although from its inception this has been regarded as an extraordinarily difficult theory, it is in fact the simplest theory to consummate the union of special relativity and Newtonian gravity. Einstein's 'popular articles' set a high standard which is now emulated by many in the range of introductory textbooks. Having mastered one of these the new reader is recommended to move next to one of the more specialized monographs, e.g. Chandrasekhar, 1983, Kramer et al., 1980, before considering review anthologies such as Einstein (centenary), Hawking and Israel, 1979, Held, 1980 and Newton (tercentenary), Hawking and Israel, 1987. As plausible gravitational wave detectors come on line in the next decade (or two) interest will focus on gravitational radiation from isolated sources, e.g., a collapsing star or a binary system including one, and I have therefore chosen to concentrate in this book on the theoretical background to this topic.

The material for the first three chapters is based on my lecture courses for graduate students. The first chapter of this book presents an account of local differential geometry for the benefit of the beginner and as a reminder of notation for more experienced readers. Chapter 2 is devoted to two-component spinors which give a representation of the Lorentz group appropriate for the description of gravitational radiation. (The relationship to the more common Dirac four-component spinors is discussed in an appendix.) Far from an isolated gravitating object one might expect spacetime to become asymptotically Minkowskian, so that the description of the gravitational field would be especially simple. This concept, *asymptotia* (asymptotic Utopia) is discussed in chapter

3, commencing with an account of the asymptotics of Minkowski spacetime, and going on to the definitions of asymptotic flatness and radiating spacetimes. (For a more detailed development of the material in chapters 2 and 3, the reader is referred to Penrose and Rindler, 1984, 1986.) The book concludes with a self-contained discussion of the characteristic initial value problem, caustics and their relation to the singularity theory of Arnol'd.

Exercises form an integral part of each chapter giving the reader a chance to check his understanding of intricate material or offering straightforward extensions of the mainstream discussions. *Problems* are even more important, for they are not only more challenging exercises, but can frequently be combined to produce significant results, encouraging the student to develop his understanding by deriving much material which is not explicitly spelt out. The brevity of this book is deceptive.

Finally I acknowledge the considerable benefit of discussions with many of my colleagues, especially Jürgen Ehlers, Bernd Schmidt and Martin Walker. In particular I owe especial thanks to Helmut Friedrich for teaching me (almost) all I know about the characteristic initial value problem.

John Stewart

1

DIFFERENTIAL GEOMETRY

The natural arena for physics is spacetime. As we shall see later spacetime is curved. It is necessary therefore to introduce a fair amount of mathematics in order to understand the physics. Fortunately we shall only need a local theory, so that problems from differential topology will not occur.

1.1 Differentiable manifolds

The simplest example of a curved space is the surface of a sphere S^2, such as the surface of the earth. One can set up local coordinates, e.g., latitude and longitude, which map S^2 onto a plane piece of paper R^2, known to sailors as a chart. Collections of charts are called atlases. Perusal of any atlas will reveal that there is no 1-1 map of S^2 into R^2; we need several charts to cover S^2. Let us state this more formally.

(1.1.1) DEFINITION

*Given a (topological) space M, a **chart on M** is a 1-1 map ϕ from an open subset $U \subset M$ to an open subset $\phi(U) \subset R^n$, i.e., a map $\phi : M \rightarrow R^n$. A chart is often called a **coordinate system.***

Now suppose the domains U_1, U_2 of two charts ϕ_1, ϕ_2 overlap in $U_1 \cap U_2$. Choose a point x_1 in $\phi_1(U_1 \cap U_2)$. It corresponds to a point p in $U_1 \cap U_2$, where $p = \phi_1^{-1}(x_1)$. Since p is in U_2 we can map it to a point x_2 in $\phi_2(U_2)$. We shall require the map $x_1 \mapsto x_2$ to be smooth, see figure 1.1.1, so that in the next section the definitions of smooth

1

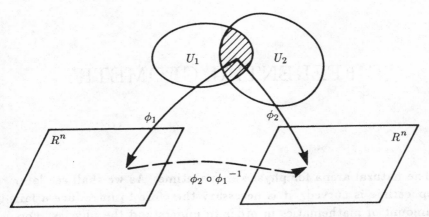

Fig. 1.1.1   When the domains of two charts ϕ_1 and ϕ_2 overlap they are required to mesh smoothly, i.e., $\phi_2 \circ \phi_1^{-1}$ must be smooth.

curves and tangent vectors will be coordinate-independent. More precisely:

(1.1.2) DEFINITION

Two charts ϕ_1, ϕ_2 are C^∞-related if both the map

$$\phi_2 \circ \phi_1^{-1} : \phi_1(U_1 \cap U_2) \to \phi_2(U_1 \cap U_2),$$

*and its inverse are C^∞. A collection of C^∞-related charts such that every point of M lies in the domain of at least one chart forms an **atlas**. The collection of all such C^∞-related charts forms a **maximal atlas**. If M is a space and A its maximal atlas, the set (M, A) is a $(C^\infty$-) **differentiable manifold**. If for each ϕ in the atlas the map $\phi : U \to R^n$ has the same n, then the manifold has **dimension** n.*

When problems of differentiablity arise we can similarly define C^k-related charts and C^k-manifolds.

The reader for whom these ideas are new is strongly recommended to peruse a geographical atlas and identify the features described above. Further examples worth examining include the plane R^2, the circle S^1, the sphere S^2 and a Möbius band.

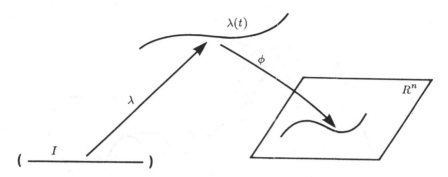

Fig. 1.2.1   A curve $\lambda : I \to M$, $t \mapsto \lambda(t)$ is smooth iff its image under a chart is a smooth curve in R^n.

1.2   Tangent vectors and tangent spaces

Most concepts in physics involve the concept of differentiability, and, as we shall see, an essential ingredient is the generalization of the idea of a vector. Simple naïve definitions of vectors do not work in general manifolds. The vector London $\to$ Paris may be parallel to the vector London $\to$ Dublin in one chart and perpendicular to it in another. Some experiments will show that there are severe problems in establishing chart-independence for the usual properties of vectors in any non-local definition. The following approach may not be an obvious one but it will capture the intuitive concepts. We start by defining a curve within a manifold.

(1.2.1) DEFINITION

A C^∞-curve in a manifold M is a map λ of the open interval $I = (a,b) \in R \to M$ such that for any chart ϕ, $\phi \circ \lambda : I \to R^n$ is a C^∞ map.

There are a number of points to note about this definition. Firstly C^k-curves are defined in the obvious way. Either or both of a, b can be infinite. By considering half-closed or closed intervals I one or both endpoints can be included. Finally the definition

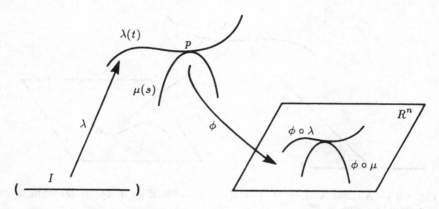

Fig. 1.2.2   Two curves $\lambda(t)$, $\mu(s)$ are tangent at p iff their images under a chart ϕ are tangent at $\phi(p)$ in R^n.

implies that the curve is parametrized. As an abbreviation one often speaks of the ***curve*** $\lambda(t)$, with $t \in (a,b)$.

Let $f : M \to R$ be a smooth function on M. Consider the map $f \circ \lambda : I \to R, t \mapsto f(\lambda(t))$. This has a well-defined derivative, the rate of change of f along the curve. Now suppose that two curves $\lambda(t), \mu(s)$ meet at a point p where $t = t_o$, $s = s_o$. Suppose that

$$\frac{d}{dt}(f \circ \lambda) = \frac{d}{ds}(f \circ \mu) \qquad \text{at} \quad t = t_o, \; s = s_o, \qquad (1.2.1)$$

for all functions f. This is a precise way of stating that "λ, μ pass through p with the same velocity". To see this we consider:

(1.2.2) LEMMA

Suppose that ϕ is any chart whose domain of dependence includes p. Let ϕ map $q \in M$ to $x^i(q)$ where $x^i(q)$ are the coordinates of q. Then (1.2.1) holds if and only if for all charts and each i

$$\left[\frac{d}{dt}(x^i \circ \lambda)\right]_{t=t_o} = \left[\frac{d}{ds}(x^i \circ \mu)\right]_{s=s_o}. \qquad (1.2.2)$$

Proof: It is trivial to show that (1.2.1) implies (1.2.2). To proceed in the other direction write $f \circ \lambda = (f \circ \phi^{-1}) \circ (\phi \circ \lambda)$. Now $f \circ \phi^{-1}$

is a map $R^n \to R$, $x^i \mapsto f(x^i) = f(\phi^{-1}(x^i))$. Also $\phi \circ \lambda$ is a map $I \to R^n$, $t \mapsto x^i(\lambda(t))$. Using the chain rule for differentiation

$$\frac{d}{dt}(f \circ \lambda) = \sum_1^n \left[\frac{\partial}{\partial x^i}(f(x^i))\right] \frac{dx^i}{dt}(\lambda(t)). \qquad (1.2.3)$$

A similar expression holds for $f \circ \mu$ which then proves the result.

Thus given a curve $\lambda(t)$ and a function f we can obtain a new number $[d(f \circ \lambda)/dt]_{t=t_o}$, the rate of change of f along the curve $\lambda(t)$ at $t = t_o$. We now use this result to remove the non-locality from the idea of a vector.

(1.2.3) DEFINITION

*The **tangent vector** $\dot{\lambda}_p = (d\lambda/dt)_p$ to a curve $\lambda(t)$ at a point p on it is the map from the set of real functions f defined in a neighbourhood of p to R, defined by*

$$\dot{\lambda}_p : f \mapsto \left[\frac{d}{dt}(f \circ \lambda)\right]_p = (f \circ \lambda)\dot{}_p = \dot{\lambda}_p(f). \qquad (1.2.4)$$

*Given a chart ϕ with coordinates x^i, **the components of $\dot{\lambda}_p$ with respect to the chart** are*

$$(x^i \circ \lambda)\dot{}_p = \left[\frac{d}{dt}x^i(\lambda(t))\right]_p.$$

*The set of tangent vectors at p is the **tangent space** $T_p(M)$ at p.*

This accords with the usual algebraists' idea of vectors, as we see from the following result.

(1.2.4) THEOREM

If the dimension of M is n then $T_p(M)$ is a vector space of dimension n.

Proof: We show first that $T_p(M)$ is a vector space, i.e., if $X_p, Y_p \in T_p(M)$ and a is a real number then

$$X_p + Y_p, \quad aX_p \quad \in T_p(M).$$

In other words we have first to show that there is a curve $\nu(t)$ through p, $t = t_o$ such that

$$\dot{\nu}_p(f) = (f \circ \nu)_p^{\cdot} = X_p f + Y_p f. \tag{1.2.5}$$

Let λ, μ be curves through p with $\lambda(t_o) = \mu(t_o) = p$, and $\dot{\lambda}_p = X_p$, $\dot{\mu}_p = Y_p$. Then $\tilde{\nu} : t \mapsto \phi \circ \lambda + \phi \circ \mu - \phi(p)$ is a curve in R^n (where $+, -$ have their usual meaning) and $\nu : t \mapsto \phi^{-1} \circ \tilde{\nu}$ is a curve in M satisfying (1.2.5). The second part of this proof is left as an exercise for the reader.

Finally we have to show that a basis of $T_p(M)$ contains n vectors. We first establish a useful result. Let ϕ be a chart with coordinates x^i. Consider n curves $\lambda_k(t)$ defined as follows

$$\phi(\lambda_k(t)) = (x^1(p), \ldots, x^{k-1}(p), x^k(p) + t, x^{k+1}(p), \ldots, x^n(p)),$$

i.e., only the k'th coordinate varies. We denote the tangent vector at p, $t = 0$ by

$$\dot{\lambda}_k(0) = \left(\frac{\partial}{\partial x^k}\right)_p. \tag{1.2.6}$$

Note the simple result $(x^i \circ \lambda_k)^{\cdot} = \delta^i{}_k$. Then using the chain rule (1.2.3) we have

$$\left(\frac{\partial}{\partial x^k}\right)_p f = \frac{d}{dt}(f \circ \lambda_k)_p$$

$$= \frac{d}{dt}[(f \circ \phi^{-1}) \circ (\phi \circ \lambda_k)]_p$$

$$= \sum_1^n \frac{\partial}{\partial x^m}(f \circ \phi^{-1}) \frac{d}{dt}(x^m \circ \lambda_k)$$

$$= \sum_1^n \frac{\partial}{\partial x^m}(f \circ \phi^{-1}) \delta_k{}^m = \left[\frac{\partial}{\partial x^k}(f \circ \phi^{-1})\right]_{\phi(p)}. \tag{1.2.7}$$

Next we show that each vector at p is a linear combination of the $(\partial/\partial x^k)_p$. To see this let X_p be a tangent vector to the curve $\lambda(t)$ with $\lambda(0) = p$. Then

$$X_p f = (f \circ \lambda)^{\cdot}(0) = [f \circ \phi^{-1} \circ \phi \circ \lambda]^{\cdot}(0)$$

$$= \sum_k \frac{\partial}{\partial x^k}(f \circ \phi^{-1})\,(x^k \circ \lambda)^{\cdot}(0)$$

$$= \sum_k \left(\frac{\partial}{\partial x^k}\right)_p f\ (X_p x^k), \qquad (1.2.8)$$

where (1.2.7) has been used in the last step. Thus

$$X_p = \sum_k (X_p x^k)\left(\frac{\partial}{\partial x^k}\right)_p, \qquad (1.2.9)$$

and so the $(\partial/\partial x^k)_p$ span $T_p(M)$. Finally to show that they are linearly independent suppose that $\sum A^k\,(\partial/\partial x^k)_p = 0$. Then

$$0 = \sum_k A^k \left(\frac{\partial}{\partial x^k}\right)_p x^i = \sum_k A^k \delta_k{}^i = A^i.$$

Thus each $A^i = 0$, i.e., the $(\partial/\partial x^k)_p$ form a basis. From (1.2.9) we see that $X_p x^i$ are the components of X_p with respect to the given basis.

We shall subsequently use the **Einstein summation convention**: in an expression where the same index occurs twice, once up once down, it is to be summed. (In a fraction up in a denominator counts as down in the numerator and vice versa.) Thus

$$\sum_k A^k \left(\frac{\partial}{\partial x^k}\right)_p \rightarrow A^k \left(\frac{\partial}{\partial x^k}\right)_p$$

WARNING: Do not confuse the differential operator acting on functions in R^n, $\partial/\partial x^k$ with the vector $(\partial/\partial x^k)_p$ in $T_p(M)$.

The final result needed to capture the concept of a tangent vector as a derivative is left as an exercise.

(1.2.5) Exercise. Suppose that f, g are two real functions in M and $fg : M \to R$ is defined by $fg(p) = f(p)g(p)$. If $X_p \in T_p(M)$ show that

$$X_p(fg) = (X_p f)g(p) + f(p)(X_p g), \qquad (1.2.10)$$

the **Leibniz product rule.**

1.3   Tensor algebra

Now that it as been demonstrated that $T_p(M)$ is a vector space it is possible to incorporate most of the machinery of linear algebra.

(1.3.1) DEFINITION

*The **dual space** $T_p^*(M)$ of $T_p(M)$ is the vector space of linear maps $\lambda : T_p(M) \to R$.*

$T_p^*(M)$ has also dimension n. If λ acts on $X_p \in T_p(M)$ the result is written either as $\lambda(X_p)$ or $< \lambda, X_p >$, a scalar. Elements of $T_p^*(M)$ are called *1-forms, forms* or *covectors*.

Let f be any function on M. For each $X_p \in T_p(M)$ $X_p f$ is a scalar. Thus f defines a map $df : T_p(M) \to R$ via

$$df(X_p) = X_p f. \qquad (1.3.1)$$

Since X_p is linear so is df and so $df \in T_p^*(M)$. df is the **differential** or **gradient** of f.

(1.3.2) LEMMA

*Let ϕ be a chart with coordinates x^i. Then the coordinate differentials dx^i form a basis for $T_p^*(M)$, the **dual basis**, with the property*

$$\left\langle dx^i, \left(\frac{\partial}{\partial x^m}\right)_p \right\rangle = \delta^i{}_m. \qquad (1.3.2)$$

Proof: Using (1.3.1) and (1.2.7) we have

$$\left\langle dx^i, \left(\frac{\partial}{\partial x^m}\right)_p \right\rangle = \left(\frac{\partial}{\partial x^m}\right)_p x^i$$

$$= \frac{\partial}{\partial x^m}(x^i \circ \phi^{-1})_{\phi(p)},$$

$$= \delta^i{}_m.$$

This proves (1.3.2). Next we show that we have a basis. Now

$$\omega_i \, dx^i = 0 \quad \Rightarrow \quad \omega_i \left\langle dx^i, \left(\frac{\partial}{\partial x^m}\right)_p \right\rangle = 0 \quad \Rightarrow \quad \omega_m = 0.$$

Thus the dx^i are linearly independent. Finally consider the form

$$\omega = \lambda - \left\langle \lambda, \left(\frac{\partial}{\partial x^k}\right)_p \right\rangle dx^k. \qquad (1.3.3)$$

Clearly

$$\left\langle \omega, \left(\frac{\partial}{\partial x^i}\right)_p \right\rangle = \left\langle \lambda, \left(\frac{\partial}{\partial x^i}\right)_p \right\rangle - \left\langle \lambda, \left(\frac{\partial}{\partial x^k}\right)_p \right\rangle \left\langle dx^k, \left(\frac{\partial}{\partial x^i}\right)_p \right\rangle$$

$$= 0,$$

for all m. Thus $\omega = 0$, and so every element of $T_p^*(M)$ is a linear combination of the dx^i.

We thus see that every 1-form can be written

$$\lambda = \left\langle \lambda, \left(\frac{\partial}{\partial x^i}\right)_p \right\rangle dx^i. \qquad (1.3.4)$$

In particular if we set $\lambda = df$

$$\left\langle df, \left(\frac{\partial}{\partial x^i}\right)_p \right\rangle = \left(\frac{\partial}{\partial x^i}\right)_p f = (f_{,i})_p,$$

where we have used (1.2.7), and the shorthand $f_{,i} = \partial f/\partial x^i$. This is the justification for the name gradient for df.

Next we need to consider the possibility of coordinate transformations $x^i \mapsto y^i(x^m)$. Using the above result

$$dy^i = A^i{}_m dx^m, \text{ where } A^i{}_m = \left(\frac{\partial y^i}{\partial x^m}\right)_p. \qquad (1.3.5)$$

This gives the result for a change of basis in $T_p^*(M)$. If $(\partial/\partial x^i)_p$ is the dual basis in $T_p(M)$, (1.3.2) implies that

$$\left(\frac{\partial}{\partial y^m}\right)_p = (A^{-1})_m{}^i \left(\frac{\partial}{\partial x^i}\right)_p, \qquad (A^{-1})_m{}^i = \left(\frac{\partial x^i}{\partial y^m}\right)_p. \quad (1.3.6)$$

The positions of the indices should be noted.

Equations (1.3.5),(1.3.6) are easily generalized. Let (e_a) be any basis of $T_p(M)$ and (ω^a) the dual basis of $T_p^*(M)$, i.e.

$$< \omega^a, e_b >= \delta^a{}_b. \qquad (1.3.7)$$

Then the formula for a change of basis $\omega^a \mapsto \widehat{\omega}^a, \, e_a \mapsto \widehat{e}_a$ is

$$\widehat{\omega}^a = A^a{}_b \omega^b, \qquad \widehat{e}_a = (A^{-1})_a{}^b e_b, \qquad (1.3.8)$$

where $A^a{}_b$ is a non-singular matrix.

Suppose the 1-form λ has components λ_a with respect to the basis ω^a, i.e.,

$$\lambda = \lambda_a \omega^a = \widehat{\lambda}_a \widehat{\omega}^a = \widehat{\lambda}_a A^a{}_b \omega^b.$$

Then since the ω^a are linearly independent

$$\lambda_b = A^a{}_b \widehat{\lambda}_a,$$

or

$$\widehat{\lambda}_a = (A^{-1})_a{}^b \lambda_b. \qquad (1.3.9)$$

Similarly if X_p is a vector, $X_p = X^a e_a = \widehat{X}^a \widehat{e}_a$ or

$$\widehat{X}^a = A^a{}_b X^b. \qquad (1.3.10)$$

Note carefully the similarities and differences in these rules. Quantities which transform like (1.3.9), (or (1.3.10)) are called **covariant** (or **contravariant**) in old-fashioned textbooks.

(1.3.3) DEFINITION

*A $(1,2)$ **tensor** S on $T_p(M)$ is a map*

$$S : T_p(M) \times T_p(M) \times T_p^*(M) \to R,$$

which is linear in each argument.

A $(1,2)$ tensor is determined by its components $S_{ab}{}^c$ with respect to a basis. Suppose $(e_a), (\omega^a)$ are dual bases. Define

$$S_{ab}{}^c = S(e_a, e_b, \omega^c) \in R.$$

Then if $X = X^a e_a$, $Y = Y^b e_b$, $\lambda = \lambda_c \omega^c$ are two vectors and a 1-form

$$
\begin{aligned}
S(X,Y,\lambda) &= S(X^a e_a, Y^b e_b, \lambda_c \omega^c) \\
&= X^a Y^b \lambda_c S(e_a, e_b, \omega^c) \qquad \text{by the linearity} \\
&= X^a Y^b \lambda_c S_{ab}{}^c.
\end{aligned}
$$

These properties generalize easily to (r,s) tensors.

Notice that two (r,s) tensors can be added and the product of a scalar and a (r,s) tensor is also a (r,s) tensor. Thus the set of (r,s) tensors forms a vector space. However the sum of a $(0,1)$ tensor, (a 1-form) and a $(1,15)$ tensor is not defined.

(1.3.4) Exercise. What is the tensor transformation law corresponding to equations (1.3.9),(1.3.10)? Suppose that the components of two tensors are equal in one basis. Show that this is then true for every other basis.

(1.3.5) Exercise. Suppose that B_{ab} has the property that whenever $C_{cd}{}^e$ is a tensor then so is $A_{abcd}{}^e = B_{ab} C_{cd}{}^e$. Prove that B_{ab} is a tensor and generalize the result to tensors of arbitrary order or **valence**.

Suppose that S is $(1,2)$ and T is $(1,1)$. Consider the map

$$U : T_p(M) \times T_p(M) \times T_p^*(M) \times T_p(M) \times T_p^*(M) \to R,$$

$$U(X,Y,\lambda,Z,\mu) = S(X,Y,\lambda)T(Z,\mu).$$

This is clearly linear in every argument and so is a $(2,3)$ tensor with components

$$U_{ab}{}^c{}_d{}^e = S_{ab}{}^c T_d{}^e.$$

There is another tensor operation to be considered which reduces the order of a tensor. Suppose U is $(2,3)$ with components $U_{ab}{}^c{}_d{}^e$. Let

$$V_{ab}{}^e = U_{ab}{}^c{}_c{}^e.$$

(1.3.6) Exercise. Show that this defines a $(1, 2)$ tensor V, the **contraction of U across the third and fourth indices.** Since this definition was given in terms of components it is necessary to verify not only linearity but also independence of the basis chosen.

Two other tensor operations will be frequently used. Suppose S is $(1, 2)$ with components $S_{ab}{}^c$. Define

$$S_{(ab)}{}^c = \tfrac{1}{2}(S_{ab}{}^c + S_{ba}{}^c), \qquad S_{[ab]}{}^c = \tfrac{1}{2}(S_{ab}{}^c - S_{ba}{}^c),$$

called **symmetrization** and **antisymmetrization** respectively. There are obvious extensions e.g.

$$T_{[abc]} = \frac{1}{3!}\left[T_{abc} + T_{bca} + T_{cab} - T_{acb} - T_{cba} - T_{bac}\right].$$

Note that these operations can be applied to subsets of the indices of higher rank tensors. A tensor which is symmetric on all of its indices is **symmetric.** A tensor which is antisymmetric on all of its indices is **antisymmetric** or **skew.**

(1.3.7) Exercise. Let M be a manifold and $f : M \to R$ a smooth function such that $df = 0$ at some point $p \in M$. Let x^i be a coordinate chart defined in a neighbourhood of p. Define

$$F_{ik} = \frac{\partial^2 f}{\partial x^i \partial x^k}.$$

By considering the transformation law for components show that F_{ik} defines a $(0, 2)$ tensor, the **Hessian of f at p.** Construct also a coordinate-free definition and demonstrate its tensorial properties. What happens if $df \neq 0$ at p?

1.4   Tensor fields and commutators

Until now we have considered only the algebra of tensors defined at one point. Next we consider the union of tangent spaces $T_p(M)$ at all points $p \in M$

$$T(M) = \bigcup_{p \in M} T_p(M).$$

(1.4.1) DEFINITION

*A **vector field on** M is a map $X : M \to T(M)$ such that $X(p) = X_p$ is a vector in $T_p(M)$.*

Thus a vector field is a specification of a vector at each point of the manifold. Given a coordinate system x^i and associated basis $(\partial/\partial x^i)_p$ for each $T_p(M)$, X has components X_p^i, where

$$X_p = X_p^i \left(\frac{\partial}{\partial x^i} \right)_p, \quad \text{and} \quad X_p^i = X_p(x^i).$$

(Here $X_p(x^i)$ means the vector X_p acting on the function x^i.) Although it is not entirely unambiguous it is conventional to drop the suffix 'p'. X is called *smooth* if the functions $X^i = X(x^i)$ are smooth for one, and hence all, charts in the atlas. Analogous definitions hold for covector and more general tensor fields.

Now suppose that X, Y are vector fields. We have already seen that for any function f, $Y_p f$ is a number for each p, the tangential derivative of f along Y at p. Thus there exists a function $Yf : M \to R$, $Yf(p) = Y_p f$. Since Yf is a function we may compute $X_p(Yf)$. In coordinates

$$X_p(Yf) = X^i \left(\frac{\partial}{\partial x^i} \right)_p Yf = X^i \left(\frac{\partial}{\partial x^i} \right)_p \left(Y^m \frac{\partial f}{\partial x^m} \right)_p$$

$$= X^i Y^m \left(\frac{\partial^2 f}{\partial x^i \partial x^m} \right)_p + X^i \left(\frac{\partial Y^m}{\partial x^i} \right)_p \left(\frac{\partial f}{\partial x^m} \right)_p.$$

This computation shows that there exists no vector Z such that $Z_p f = X_p(Yf)$. For if there were then

$$X_p(Yf) = Z_p f = Z^i \left(\frac{\partial f}{\partial x^i} \right)_p.$$

But we have just seen that $X_p(Yf)$ involves second derivatives of f. However consider the possibility

$$Z_p f = X_p(Yf) - Y_p(Xf) \quad \text{for all } f.$$

The right hand side is

$$X^i Y^m{}_{,i} \left(\frac{\partial f}{\partial x^m} \right)_p - Y^i X^m{}_{,i} \left(\frac{\partial f}{\partial x^m} \right)_p.$$

Thus X, Y define a new vector field, the **commutator** $[X, Y]$ such that

$$[X, Y]_p f = X_p(Yf) - Y_p(Xf) \qquad \text{for all } f.$$

It has components

$$[X, Y]^i = X^m Y^i{}_{,m} - Y^m X^i{}_{,m}.$$

The commutator has an important geometrical interpretation described in section 1.6.

(1.4.2) Exercise. Show directly from the definition that

$$[X, Y] = -[Y, X], \qquad [X + Y, Z] = [X, Z] + [Y, Z],$$

and prove the **Jacobi identity**

$$[[X, Y], Z] + [[Y, Z], X] + [[Z, X], Y] = 0.$$

1.5 Maps of manifolds

In this section we consider maps from one manifold M to another N. In the most important applications M and N are the same manifold.

(1.5.1) DEFINITION

A map $h : M \to N$ is C^∞ if for every C^∞ function $f : N \to R$ the function $fh : M \to R$ is also C^∞.

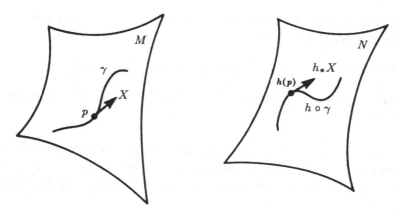

Fig. 1.5.1 $h : M \to N$ maps curves in M to curves in N. The push-forward h_* maps tangent vectors of curves in M to tangent vectors of curves in N.

Now suppose that $h : M \to N$ is C^∞. Let γ be a C^∞ curve in M. Then $h(\gamma)$ is a C^∞ curve in N and h can be used to map tangent vectors over M to tangent vectors over N.

(1.5.2) DEFINITION

*Given a C^∞ map $h : M \to N$ there is a map $h_* : T_p(M) \to T_{h(p)}(N)$, the **induced linear map**, or **push-forward** which maps the tangent vector of a curve γ at $p \in M$ to the tangent vector of the curve $h(\gamma)$ at $h(p) \in N$.*

Of course this definition, while highly intuitive, is of little use for computational purposes. For those we need a further result:

(1.5.3) LEMMA

For each C^∞ function $f : N \to R$

$$(h_* X_p)f = X_p(f \circ h),\tag{1.5.1}$$

and h_* is a linear map.

Proof: Let γ be a curve through p with $\gamma(0) = p, \dot{\gamma}(0) = X_p$. From the definition

$$(h_*X_p)f = [f \circ (h \circ \gamma)]\dot{}(0) = [(f \circ h) \circ \gamma]\dot{}(0) = X_p(f \circ h).$$

Furthermore the right hand side is linear in X_p and so therefore must the left hand side be.

Once we have a map between vectors on the two manifolds we can construct a map between covectors on the two manifolds. There is however a slight surprise; the map is in the opposite direction!

(1.5.4) DEFINITION

The maps $h : M \rightarrow N$, $h_* : T_p(M) \rightarrow T_{h(p)}(N)$ *induce a linear map, the* **pull-back** $h^* : T^*_{h(p)}(N) \rightarrow T^*_p(M)$ *as follows: if* $\omega \in T^*_{h(p)}(N)$, $X_p \in T_p(M)$ *then*

$$(h^*\omega)(X_p) = \omega(h_*X_p). \tag{1.5.2}$$

Let us now consider a concrete example in order to show how equations (1.5.1),(1.5.2) yield explicit formulae for the maps h_* and h^*. Let M be R^3 and N be R^2 with coordinate systems (x, y, z), (x, y) respectively. Let $h(x, y, z) = (x, y)$. Then for any $f : N \rightarrow R$

$$f \circ h(x, y, z) = f(x, y).$$

If X_p has components (X, Y, Z), and g is a function on M

$$X_p g = \left(X\frac{\partial g}{\partial x} + Y\frac{\partial g}{\partial y} + Z\frac{\partial g}{\partial z} \right)_p.$$

Thus

$$X_p(f \circ h) = \left(X\frac{\partial f}{\partial x} + Y\frac{\partial f}{\partial y} \right)_p = (h_*X_p)f.$$

We see that $h_*X_p \in T_{h(p)}(N)$ has components (X, Y). Finally if $\omega \in T^*_{h(p)}(N)$ has components (λ, μ), i.e., $\omega = \lambda dx + \mu dy$, then

$$\omega(h_*X_p) = \lambda X + \mu Y = h^*\omega(X_p).$$

Thus $h^*\omega \in T_p^*(M)$ has components $(\lambda, \mu, 0)$, i.e.

$$h^*\omega = \lambda dx + \mu dy + 0dz.$$

(1.5.5) Exercise. Carry out the corresponding calculation when $M = R^2$ and $N = R^3$.

(1.5.6) Exercise. Let $h : M \to N$ be a smooth map between two manifolds M and N. We may extend the definition of h^* to map functions $f : N \to R$ to functions $h^*f : M \to R$ via

$$h^*f = f \circ h.$$

Show that h^* commutes with the differential operator, i.e.

$$h^*(df) = d(h^*f).$$

1.6   Integral curves and Lie derivatives

Given a vector field X, an *integral curve of X in M* is a curve γ in M such that at each point p on γ, the tangent vector is X_p. Existence and uniqueness of such curves is guaranteed by:

(1.6.1) THEOREM

Every smooth vector field X defines locally a unique integral curve γ through each point p such that $\gamma(0) = p$.

Proof: Let x^i be a local coordinate chart and let x_p^i be the coordinates of p. The equation of the integral curve is

$$\frac{d}{dt} x^i(t) = X^i(x^m(t)),$$

with the initial conditions $x^i(0) = x_p^i$. Provided X^i is smooth the standard theory of ordinary differential equations guarantees the existence, at least locally, i.e., for small t, of precisely one solution.

A very useful result for what follows is:

(1.6.2) LEMMA

If X is a smooth vector field such that $X_p \neq 0$, then there exists a coordinate system y^i defined in a neighbourhood U of p such that in U, $X = \partial/\partial y^1$.

Proof: (We give the proof in 4 dimensions; the generalization is obvious.) By continuity there exists some neighbourhood U of p, in which $X \neq 0$. We choose a 3-surface Σ in U nowhere tangent to X, with coordinates (y^2, y^3, y^4). There is a unique integral curve of X through each point of Σ. Define the function y^1 along each such curve by $y^1 = \int dt$, with $y^1 = 0$ on Σ, and the functions y^2, y^3, y^4 to be constant along the curves. Then (y^i) is the required coordinate system.

An integral curve $\gamma(t)$ is ***complete*** if it is defined for all values of t. A set of complete integral curves is a ***congruence***.

A smooth vector field X defines a transformation $h_s : M \to M$ for each real s as follows. For each $p \in M$, $h_s(p)$ is the point on the unique integral curve of X through p at a parameter "distance" s from p, i.e., if $\gamma(t)$ is the curve and p is $\gamma(t_o)$ then $h_s(p)$ is $\gamma(t_o+s)$. It should be obvious that as s varies these transformations form a 1-parameter Abelian group, for

$$h_{t+s} = h_{s+t}, \qquad h_0 = \text{identity}, \qquad (h_s)^{-1} = h_{-s}.$$

It is now possible to give the geometric interpretation of the commutator mentioned in section 1.4.

(1.6.3) THEOREM

Let X, Y be two smooth vector fields in M inducing groups $(h_s), (k_t)$. Let $p \in M$, and define $q = k_{dt}(p), r = h_{ds}(p), u = h_{ds}(q), v = k_{dt}(r)$. Then if (m_t) is the group induced by $[X, Y]$

$$v = m_{dsdt}(u) + O(ds^3, dt^3, \dots)$$

in the sense that this equation holds true for any coordinate system in the atlas.

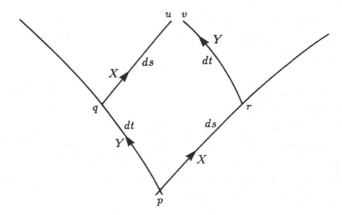

Fig. 1.6.1 The geometrical interpretation of the commutator. Moving ds in the direction X to r followed by dt in the direction Y to v is not the same as moving dt along Y to q followed by ds along X to u. v can be reached from u by moving $ds\,dt$ along the commutator.

Proof: Introduce a local coordinate system x^i with p at $x^i = 0$. Then considering integral curves of X

$$x_r^i = x_p^i + \left(\frac{\partial x^i}{\partial s}\right)_p ds + \tfrac{1}{2}\left(\frac{\partial^2 x^i}{\partial s^2}\right)_p ds^2 + \cdots$$

$$= 0 + X_p^i\, ds + \tfrac{1}{2}\left(\frac{\partial X^i}{\partial s}\right)_p ds^2 + \cdots$$

$$= X_p^i\, ds + \tfrac{1}{2}(X^i{}_{,m}X^m)_p\, ds^2 + \cdots,$$

where $\cdots$ denotes third order quantities. Similarly

$$x_v^i = x_r^i + Y_r^i\, dt + \tfrac{1}{2}(Y^i{}_{,m}Y^m)_r\, dt^2 + \cdots$$

$$= x_r^i + \left[Y_p^i + (Y^i{}_{,m}X^m)_p\, ds\right] dt + \tfrac{1}{2}(Y^i{}_{,m}Y^m)_p\, dt^2 + \cdots$$

$$= X_p^i\, ds + Y_p^i\, dt + \tfrac{1}{2}(X^i{}_{,m}X^m)_p\, ds^2 +$$
$$(Y^i{}_{,m}X^m)_p\, ds\, dt + \tfrac{1}{2}(Y^i{}_{,m}Y^m)_p\, dt^2 + \cdots.$$

The corresponding formula for x_u^i is obtained by interchanging X and Y and s and t. Thus

$$x_v^i - x_u^i = (Y^i{}_{,m}X^m - X^i{}_{,m}Y^m)_p\, ds\, dt + \cdots$$
$$= [X, Y]_p^i\, ds\, dt + \cdots.$$

To this order we can evaluate the commutator at u instead which proves the result. Thus $pquv$ forms a small quadrilateral iff the commutator vanishes.

At the moment although we can differentiate functions we cannot differentiate vectors or covectors. Differentiation implies a comparison between values at two spacetime points and we cannot do this for vectors, because the two vectors will lie in different vector spaces. However given an additional smooth vector field X there is a comparison which we can make, leading to the definition of a derivative, the Lie derivative.

(1.6.4) DEFINITION

*Let p be a point of M and let X be a smooth vector field. Let γ be the integral curve of X through p inducing a 1-parameter group of transformations (h_t). Then if $f : M \to R$ is any real function on M, the **Lie derivative of f with respect to X** is*

$$(\pounds_X f)_p = \lim_{dt \to 0} \left[\frac{f(h_{dt}(p)) - f(p)}{dt} \right].$$

This definition, while exhibiting manifest coordinate invariance, is of little use for computations. We therefore establish:

(1.6.5) THEOREM

For any function f on M,

$$(\pounds_X f)_p = X_p f. \tag{1.6.1}$$

Proof: We use the coordinate system adapted to the integral curves set up in lemma 1.6.2. The coordinates of $h_t(p)$ are

$$y^1(h_t(p)) = y^1(p) + t, \qquad y^\alpha(h_t(p)) = y^\alpha(p) \qquad \alpha = 2, 3, \ldots.$$

Then from the definition

$$(\pounds_X f)_p = \left(\frac{\partial f}{\partial y^1} \right)_p$$

$$= X_p f,$$

where lemma 1.6.2 has been used. Thus equation (1.6.1) is true in this coordinate system. But being a scalar equation it is true in all coordinate systems. (C.f. exercise 1.3.4)

Now let $q = h_{-dt}(p)$ so that $p = h_{dt}(q)$. Using the results of section 1.5 we have an induced linear map $(h_{dt})_*$ mapping $T_q(M)$ to $T_p(M)$, and so we may compare vectors at q with vectors at p. This leads to the notion of a derivative for vectors.

(1.6.6) DEFINITION

*Using the notation of definition 1.6.4 the **Lie derivative of a smooth vector field** Z **with respect to** X **at** p **is***

$$(\pounds_X Z)_p = \lim_{dt \to 0} \left[\frac{Z_p - (h_{dt})_* Z_{h_{-dt}(p)}}{dt} \right],$$

a well defined vector at p.

Once more we need to establish a suitable formula for computations.

(1.6.7) THEOREM

$$(\pounds_X Z)_p = [X, Z]_p. \tag{1.6.2}$$

Proof: Again we use the special coordinate system adapted to the congruence. Then, letting $\alpha = 2, 3, \ldots$

$$(\pounds_X Z)_p^i = \lim_{dt \to 0} \left[\frac{Z^i(y_p^1, y_p^\alpha) - Z^i(y_p^1 - dt, y_p^\alpha)}{dt} \right]$$

$$= \left(\frac{\partial Z^i}{\partial y^1} \right)_p = X_p(Z^i).$$

Now the right hand side is not a vector; see, e.g., the discussion in section 1.4. However suppose that we subtract $Z_p(X^i)$ from the right hand side. (Since $X^i = \delta^i{}_1$ this is zero and we are free to do it.) Then

$$(\pounds_X Z)^i_p = X_p(Z^i) - Z_p(X^i) = [X,Z]^i_p.$$

Now both sides are vectors and so the equation holds in all coordinate systems since it holds in our special one. (C.f. exercise 1.3.4 again.)

We can now use the induced linear map $(h_{dt})^*$ to define Lie derivatives of covectors. In practice we can use a short cut, based on the requirement that a derivative must obey the Leibniz rule. Thus let ω be a covector. Then for any vector field Y, $\omega(Y)$ is a function. Therefore in an arbitrary coordinate system

$$\begin{aligned}
\pounds_X(\omega(Y)) &= X(\omega(Y)) \\
&= (\omega(Y))_{,i} X^i \\
&= (\omega_k Y^k)_{,i} X^i \\
&= \omega_{k,i} Y^k X^i + \omega_k Y^k{}_{,i} X^i.
\end{aligned}$$

However we must require the Leibniz rule to hold, i.e.

$$\begin{aligned}
\pounds_X(\omega(Y)) &= (\pounds_X \omega)(Y) + \omega(\pounds_X Y) \\
&= (\pounds_X \omega)_k Y^k + \omega_k(\pounds_X Y)^k \\
&= (\pounds_X \omega)_k Y^k + \omega_k Y^k{}_{,i} X^i - \omega_k X^k{}_{,i} Y^i.
\end{aligned}$$

Thus

$$(\pounds_X \omega)_k Y^k = \omega_{k,i} Y^k X^i + \omega_k X^k{}_{,i} Y^i.$$

This has to hold for all Y and so we see that the local components of the covector $\pounds_X \omega$ are

$$(\pounds_X \omega)_k = \omega_{k,i} X^i + \omega_i X^i{}_{,k}. \tag{1.6.3}$$

(1.6.8) Exercise. Write out a formal definition along the lines of definition 1.6.6 and develop the equivalent theorem to verify equation (1.6.3).

(1.6.9) Exercise. Compute the components of $\pounds_X T$ where T is a $(1,1)$ tensor.

(1.6.10) Exercise. If X, Y, Z are smooth vector fields show that

$$\mathcal{L}_{[X,Y]}Z = \mathcal{L}_X\mathcal{L}_Y Z - \mathcal{L}_Y\mathcal{L}_X Z.$$

[Hint: at last a use for the Jacobi identity!]

(1.6.11) Exercise. X is a vector field on a manifold M and $f :$ $M \to R$ is a function. Show that Lie differentiation commutes with the differential operator, i.e.

$$\mathcal{L}_X df = d\left(\mathcal{L}_X f\right).$$

Demonstrate also that Lie differentiation commutes with (anti-) symmetrization and contraction of indices.

(1.6.12) Exercise. Let x^i, $i = 1, 2$ be a Cartesian coordinate chart on R^2. The vector field X has coordinate-induced components $X^i = (x^2, x^1)$. Construct the integral curves of X and sketch them. Let η be the $(0, 2)$ tensor field with components $\mathrm{diag}(1, -1)$. Show that

$$\mathcal{L}_X \eta = 0.$$

The finite transformations induced by translation along the integral curves of X are said to *preserve* η. In the context of special relativity they are called *boosts* or *Lorentz transformations*.

(1.6.13) Exercise. Let x^i be a Cartesian coordinate chart on R^4, and let $\eta_{ik} = \mathrm{diag}(1, -1, -1, -1)$ be the usual Minkowski metric of special relativity. For the purposes of this exercise η may be assumed to be a $(0, 2)$ tensor. Let X be a vector field on M with associated 1-parameter family of transformations h_t. Show that X preserves η iff it satisfies *Killing's equation*

$$X_{(i,m)} \equiv \eta_{k(i}X^k{}_{,m)} = 0.$$

Assuming for the moment (or referring to exercise 1.10.12) that this implies that each component X^i is a linear function of the x^k, derive the general solution, and hence obtain the *Poincaré invariance group* of special relativity.

The Lie derivative is one of the most important and least understood concepts in mathematical physics. As a final example in

this section we consider linearized perturbation theory on manifolds. In what follows $M = M_0$ is a 4-dimensional manifold, but
4 can be changed to n with only minor arithmetic changes. Let
N be a 5-dimensional smooth manifold containing a 1-parameter
family of smooth non-intersecting 4-manifolds M_ϵ. The manifolds M_ϵ are to be thought of as perturbations of a specific
one, say M_0. If one manifold is a replica of another we need a
point-identification-map which relates points in the two manifolds which are to be regarded as the "same". We supply this by
introducing a vector field V on N which is everywhere transverse
i.e., non-parallel to the M_ϵ. Points in the various M_ϵ which lie on
the same integral curve of V are to be regarded as the same. The
scaling of V is chosen so that the induced 1-parameter group of
transformations h_ϵ maps M_0 to M_ϵ. In old-fashioned notation a
choice of V is a *choice of gauge*.

Now let Q_0 be some geometric quantity on M and Q_ϵ the corresponding quantity on M_ϵ. This defines a field Q on N. Along
each integral curve of V we have a Taylor series

$$Q_\epsilon = h_{\epsilon *} \left[Q_0 + \epsilon \left(\mathcal{L}_V Q_\epsilon \right)_{\epsilon=0} + O(\epsilon^2) \right],$$

where ϵ is small. The first order term is usually called the *linearized perturbation of Q*. In general Q will satisfy some
complicated nonlinear equation, but if all quantities are expanded
as Taylor series in ϵ and nonlinear terms are discarded, the result-
ing linear equations for the perturbation may be both simple to
solve and relevant. There is however a problem, because there is
no preferred choice of V.

(1.6.14) Problem. By considering the difference between two
choices for V show that the linearized perturbation of Q is
gauge-invariant iff

$$\mathcal{L}_X Q_0 = 0,$$

for all 4-vectors X on M_0. Verify that this is the case if one of the
following holds:

i) Q_0 vanishes identically,
ii) Q_0 is a constant scalar field,

iii) Q_0 is a linear combination of products of Kronecker deltas with constant coefficients.

1.7   Linear connections

The concepts of parallelism and differentiation of vectors are crucial to classical mechanics. How are we to introduce them in a curved manifold? The Lie derivative is not sufficiently general since it depends on the choice of a vector field X. It turns out that the clearest way to introduce the new structures is axiomatically.

(1.7.1) DEFINITION

*A **linear connection** ∇ on M is a map sending every pair of smooth vector fields (X, Y) to a vector field $\nabla_X Y$ such that*

$$\nabla_X(aY + Z) = a\nabla_X Y + \nabla_X Z,$$

*for any **constant** scalar a, but*

$$\nabla_X(fY) = f\nabla_X Y + (Xf)Y,$$

when f is a function, and it is linear in X

$$\nabla_{X+fY} Z = (\nabla_X Z + f\nabla_Y Z).$$

Further, acting on functions f, ∇_X is defined by

$$\nabla_X f = Xf.$$

$\nabla_X Y$ *is called the **covariant derivative of** Y **with respect to** X. Because $\nabla_X Y$ is not linear in Y, ∇ is not a tensor. However unlike the Lie derivative $\pounds_X Y$, $\nabla_X Y$ is linear in X, thus defining a $(1,1)$ tensor ∇Y mapping X to $\nabla_X Y$, called the **covariant derivative of** Y.*

Let (e_a) be a basis for vector fields and write ∇_{e_a} as ∇_a. Since $\nabla_a e_b$ is a vector there exist scalars $\Gamma^c{}_{ba}$ such that

$$\nabla_a e_b = \Gamma^c{}_{ba} e_c. \qquad (1.7.1)$$

(Note carefully the (conventional) position of the indices!) The $\Gamma^c{}_{ba}$ are called the **components of the connection.** Now suppose that $X = X^a e_a$ etc. Then, using definition (1.7.1)

$$
\begin{aligned}
\nabla_X Y &= \nabla_{X^a e_a}(Y^b e_b) \\
&= X^a \nabla_a (Y^b e_b) \\
&= X(Y^b)e_b + X^a \nabla_a e_b Y^b \\
&= [X(Y^c) + \Gamma^c{}_{ba} X^a Y^b] \, e_c. \qquad (1.7.2)
\end{aligned}
$$

Thus $\nabla_X Y$ is completely specified by giving the components of the connection. We may also write (1.7.2) as

$$
\begin{aligned}
(\nabla_X Y)^c &= [e_a(Y^c) + \Gamma^c{}_{ba} Y^b] \, X^a \\
&= Y^c{}_{;a} X^a \quad \text{say.} \qquad (1.7.3)
\end{aligned}
$$

$Y^c{}_{;a}$ are the components of the $(1,1)$ tensor ∇Y. Neither of the two terms in $Y^c{}_{;a}$ transforms like the components of a tensor, but their sum does. In a coordinate induced basis

$$Y^i{}_{;k} = Y^i{}_{,k} + \Gamma^i{}_{mk} Y^m. \qquad (1.7.4)$$

Before we give an example we need to introduce a new concept. In Euclidean space if we move a vector without changing its magnitude or direction it is said to be parallelly transported. This can be generalized as follows:

(1.7.2) DEFINITION

If $\nabla_X Y = 0$ then Y is said to be **parallelly transported with respect to** X.

We now consider a concrete example. Let M be R^2 with Cartesian coordinates (x^1, x^2), and $e_i = \partial/\partial x^i$. The normal geometric concept of parallelism suggests that e_i is parallelly transported along e_m and so

$$\nabla_i e_m = 0, \qquad \Gamma^i{}_{mk} = 0.$$

Next consider polar coordinates (r, θ), $e_r = \partial/\partial r$, $e_\theta = r^{-1} \partial/\partial \theta$. These basis vectors are parallelly transported along e_r and so

$$\nabla_r e_r = \nabla_r e_\theta = 0.$$

They change direction when transported along e_θ. In fact a result from elementary vector calculus shows that

$$\nabla_\theta e_r = r^{-1} e_\theta, \qquad \nabla_\theta e_\theta = -r^{-1} e_r.$$

Thus all of the connection coefficients vanish except

$$\Gamma^\theta{}_{r\theta} = \frac{1}{r}, \qquad \Gamma^r{}_{\theta\theta} = -\frac{1}{r}.$$

This example illustrates the fact that the connection components cannot be the components of a tensor. For if they were they would vanish in all frames if they vanished in one. The general transformation law is given by the following theorem.

(1.7.3) THEOREM

Consider a change of basis, $\hat{e}_a = B_a{}^b e_b$. *Then*

$$\hat{\Gamma}^a{}_{bc} = (B^{-1})^a{}_f B_b{}^g B_c{}^h \Gamma^f{}_{gh} + (B^{-1})^a{}_f B_c{}^h e_h(B_b{}^f). \qquad (1.7.5)$$

Proof:

$$\begin{aligned}
\nabla_{\hat{c}} \hat{e}_b &= \hat{\Gamma}^a{}_{bc} \hat{e}_a = \hat{\Gamma}^a{}_{bc} B_a{}^f e_f \\
&= \nabla_{(B_c{}^h e_h)} (B_b{}^g e_g) \\
&= B_c{}^h B_b{}^g \nabla_h e_g + B_c{}^h (e_h B_b{}^g) e_g \\
&= B_b{}^g B_c{}^h \Gamma^f{}_{gh} e_f + B_c{}^h e_h(B_b{}^f) e_f,
\end{aligned}$$

which proves (1.7.5). The first term on the right corresponds to
the usual tensor transformation law.

(1.7.4) Exercise. Consider the example above setting $\hat{e}_1 = e_r$,
$\hat{e}_2 = e_\theta$ and compute $\hat{\Gamma}^a{}_{bc}$ explicitly using formula (1.7.5).

(1.7.5) Exercise. Suppose that ∇, ∇^* are two linear connections
on M. Show from definition 1.7.1 that

$$D(X,Y) = (\nabla_X Y - \nabla^*_X Y),$$

is a tensor. Compute also the components from (1.7.5) and show
that they transform correctly.

We can now extend ∇_X to tensors via the Leibniz rule. Suppose
that η is a covector. Then $\eta(Y)$ is a function for each vector Y.
Hence $\nabla_X(\eta(Y)) = X(\eta(Y))$. But from the Leibniz rule

$$\nabla_X(\eta(Y)) = (\nabla_X \eta)(Y) + \eta(\nabla_X Y),$$

and so $\nabla_X \eta$ is defined by

$$(\nabla_X \eta)(Y) = X(\eta(Y)) - \eta(\nabla_X Y) \qquad (1.7.6)$$

for all vectors Y. We can evaluate this in component form by
setting $X = e_a$, $Y = Y^b e_b$, $\eta = \eta_c \omega^c$, where (ω^c) is the dual basis.
Then (1.7.6) becomes

$$(\nabla_a \eta)_b Y^b = e_a(Y^b \eta_b) - \eta_c (\nabla_a(Y^b e_b))^c,$$

or

$$(\nabla_a \eta)_b = e_a(\eta_b) - \Gamma^c{}_{ba} \eta_c. \qquad (1.7.7)$$

In particular

$$\nabla_a \omega^b = -\Gamma^b{}_{ca} \omega^c, \qquad (1.7.8)$$

and for a coordinate basis

$$(\nabla_i \eta)_k \equiv \eta_{k;i} = \eta_{k,i} - \Gamma^m{}_{ki} \eta_m. \qquad (1.7.9)$$

(1.7.6) Exercise. Extend the definition of ∇ to tensors. E.g., show that if T is a $(1,1)$ tensor

$$(\nabla_a T)^c{}_b = e_a(T^c{}_b) + \Gamma^c{}_{fa}T^f{}_b - \Gamma^d{}_{ba}T^c{}_d, \qquad (1.7.10)$$

and generalize this result to arbitrary tensors.

1.8   Geodesics

In flat space a geodesic is the shortest distance between two points, i.e., a straight line. It has the property that its tangent vector is parallelly transported along itself; if the tangent vector is X then $\nabla_X X = 0$. We now generalize this concept to a manifold.

(1.8.1) DEFINITION

Let X be a vector field such that $\nabla_X X = 0$. Then the integral curves of X are called **geodesics**.

(1.8.2) THEOREM

There is precisely one geodesic through a given point $p \in M$ in a given direction X_p.

Proof: Choose local coordinates x^i. Let the tangent vector to some curve be X^i. Then the curve is given by

$$\frac{d}{dt}x^i(t) = X^i\left(x^k(t)\right), \qquad x^i(0) = x^i_p.$$

The curve is geodesic iff

$$0 = (\nabla_X X)^i = X^i{}_{;k}X^k = X^i{}_{,k}X^k + \Gamma^i{}_{km}X^k X^m.$$

But $f_{,i}X^i = Xf = df/dt$ and so

$$\frac{d}{dt}X^i = -\Gamma^i{}_{km}X^k X^m,$$

or

$$\frac{d^2}{dt^2} x^i(t) = -\Gamma^i{}_{km}\left(x(t)\right)\left(\frac{dx^k}{dt}\right)\left(\frac{dx^m}{dt}\right), \qquad (1.8.1)$$

which has to be solved subject to the initial data

$$x^i(0) = x^i_p, \qquad \left(\frac{dx^i}{dt}\right)_0 = X^i_p.$$

By the standard theory of ordinary differential equations the solution exists locally, i.e., for small values of t, and is unique.

Corollary. Locally we can always solve (1.8.1) subject to the 2-point boundary conditions $x^i(0) = x^i_p$, $x^i(1) = x^i_q$ where p and q are neighbouring points. Thus locally there is a unique geodesic joining two neighbouring points. In flat space this statement is true globally but in a curved space it fails. For example, the geodesics on a sphere are great circles which intersect in pairs of points.

In a local coordinate system the equation of a geodesic is (1.8.1)

$$\ddot{x}^i + \Gamma^i{}_{km}\dot{x}^k\dot{x}^m = 0,$$

where $\dot{x}^i = dx^i/dt$ etc. We now investigate the freedom in the parameter t. Suppose that it is changed to $s = s(t)$; what choices of s preserve the form (1.8.1)? Letting $x'^i = dx^i/ds$ etc., we have

$$\left(x''^i + \Gamma^i{}_{km}x'^k x'^m\right)\dot{s}^2 = -\ddot{s}x'^i. \qquad (1.8.2)$$

Thus the standard form of the equation is preserved iff $\ddot{s} = 0$, i.e.,

$$s = at + b, \qquad a, b \quad \text{constants.}$$

A parameter which produces the standard form (1.8.1) for the geodesic equation is an **affine parameter**. An affine parameter is defined up to a change of scale and origin.

(1.8.3) DEFINITION

*Let $p \in M$ be fixed. The **exponential map at** p sends each vector X_p at p to the point unit parameter distance along the unique* *geodesic through p with tangent vector X_p. In a small neighbour-hood of p the exponential map has an inverse, i.e., there exists a neighbourhood U of p such that $q \in U$ implies $q = exp(X_p)$ for some $X_p \in T_p(M)$. Now let (e_a) be a basis for $T_p(M)$, and write $q = exp(X_p^a e_a)$. The (X_p^a) are called **normal coordinates of q** and U is a **normal neighbourhood**.*

Consider the geodesic through p, q. Since the tangent vector is X_p the normal coordinates of any point r on the geodesic must be proportional to X_p^a. Thus the geodesic has the normal coordinate form

$$x^a = s X_p^a,$$

where s is some parameter. If t is an affine parameter we have $\dot{s}(0) \neq 0$. Now from (1.8.2) the equation of the geodesic is

$$\dot{s}^2 \Gamma^i_{km} X_p^k X_p^m = -\ddot{s} X_p^i,$$

and since $\dot{s}(0) \neq 0$

$$(\Gamma^i_{km})_p X_p^k X_p^m = -\left(\frac{\ddot{s}}{\dot{s}^2}\right)_0 X_p^i.$$

Since this has to hold for all directions we may deduce that

$$\Gamma^i_{(km)} = 0 \text{ at } p. \qquad\qquad (1.8.3)$$

Thus we can always find coordinates, normal coordinates, for which the symmetric part of each connection coefficient vanishes at a point. This desirable property (1.8.3) would not be possible if the connection coefficients were components of a tensor.

According to Newton's laws a body moving under the action of no forces moves along a straight line. However if gravitation is taken into account there are no force-free massive bodies. Ein-stein proposed treating gravitation as a geometrical quantity as follows: a body moving solely under the attraction of gravity and no other forces moves along a geodesic. Let $\mathbf{x} = x^\alpha = (x^1, x^2, x^3)$ be spatial coordinates and define $x^i = (t, x^\alpha)$. Then the 4-velocity of a particle is $v^i = dx^i/dt = (1, \mathbf{v})$, where $\mathbf{v}$ is the 3-velocity. For a body moving in a gravitational field with Newtonian potential ϕ

$$\ddot{x}^i = (0, -\nabla\phi) = (0, -\phi_{,\alpha}), \qquad \alpha = 1, 2, 3.$$

Thus

$$\ddot{x}^0 = 0, \qquad \ddot{x}^\alpha + \phi_{,\alpha}\dot{x}^0\dot{x}^0 = 0,$$

from which we may read off the **Newtonian connection** as

$$\Gamma^\alpha{}_{00} = \phi_{,\alpha}, \qquad \Gamma^i{}_{km} = 0 \qquad \text{otherwise.} \tag{1.8.4}$$

(1.8.4) Exercise. What is the physical significance of normal coordinates? Can the gravitational force be considered as a spacetime vector?

1.9   Torsion and curvature

Although ∇ is not a tensor there are two tensors closely related to the linear connection.

(1.9.1) DEFINITION

*The **torsion tensor** is a $(1,2)$ tensor field T defined by*

$$T(X,Y) = (\nabla_X Y - \nabla_Y X - [X,Y]), \tag{1.9.1}$$

for smooth vector fields X, Y.

This definition is not entirely transparent in that it is not obvious that T is a tensor. We should therefore exhibit linearity in the arguments. Since $T(X,Y) = -T(Y,X)$ it is only necessary to check the first argument. Trivially

$$T(X + Z, Y) = T(X, Y) + T(Z, Y).$$

Now if f is a function

$$\begin{aligned}
T_p(fX,Y) &= \left(\nabla_{fX}Y - \nabla_Y(fX) - [fX,Y]\right)_p \\
&= (f\nabla_X Y - f\nabla_Y X - (Yf)X - f[X,Y] + (Yf)X)_p \\
&= f(p)\,T_p(X,Y).
\end{aligned}$$

To compute the components of T we introduce a basis (e_a) of $T_p(M)$. Since the commutator of two vectors is a vector, there exist scalars $\gamma^a{}_{bc} = -\gamma^a{}_{cb}$ such that

$$[e_b, e_c] = \gamma^a{}_{bc} e_a. \qquad (1.9.2)$$

The $\gamma^a{}_{bc}$ are called the **commutator coefficients.**

(1.9.2) Exercise. If the basis is coordinate induced, i.e., $e_i = \partial/\partial x^i$ show that $\gamma^i{}_{km} = 0$. Verify also that in the example in R^2 discussed in section 1.7

$$\gamma^\theta{}_{r\theta} = -\gamma^\theta{}_{\theta r} = -\frac{1}{r}.$$

Continuing with our computation of the components of T we set $X = e_a$, $Y = e_b$ in (1.9.1) obtaining

$$T(e_a, e_b) = \nabla_a e_b - \nabla_b e_a - [e_a, e_b]$$
$$= [\Gamma^c{}_{ba} - \Gamma^c{}_{ab} - \gamma^c{}_{ab}] e_c,$$

so that T has components

$$T^c{}_{ab} = -2\Gamma^c{}_{[ab]} - \gamma^c{}_{ab}.$$

We have already seen that the geodesics fix only the symmetric part of the connection. Thus the torsion represents that part of the connection not determined by the geodesics. Since the physical viewpoint taken here is that the connection is introduced to describe certain curves as geodesics, we shall assume that the torsion vanishes identically. Some non-standard theories, notably Einstein-Cartan theory attempt to attach physical significance to the torsion.

The second tensor related to ∇ is considerably more important for gravitation.

(1.9.3) DEFINITION

*The **Riemann curvature tensor** is a $(1,3)$ tensor field defined by*

$$R(X, Y)Z = \nabla_Y \nabla_X Z - \nabla_X \nabla_Y Z + \nabla_{[X,Y]}Z, \qquad (1.9.3)$$

at each point. (It should be noted that some authors differ by an overall sign here.)

This is another oblique definition in which we have to exhibit linearity in the 3 arguments. However since $R(X,Y)Z = -R(Y,X)Z$ it is only necessary to verify linearity in X,Z. This is obvious for addition and multiplication by a constant. Thus we need only consider the effect of multiplying the arguments X and Z by a function f. In the first case we have

$$R(fX,Y)Z = \nabla_Y\nabla_{fX}Z - \nabla_{fX}\nabla_Y Z + \nabla_{[fX,Y]}Z$$
$$= f\nabla_Y\nabla_X Z - f\nabla_X\nabla_Y Z + (Yf)\nabla_X Z +$$
$$\nabla_{(f[X,Y]-(Yf)X)}Z$$
$$= fR(X,Y)Z.$$

In the second case we find

$$R(X,Y)fZ = \nabla_Y\nabla_X(fZ) - \nabla_X\nabla_Y(fZ) + \nabla_{[X,Y]}(fZ)$$
$$= \nabla_Y(f\nabla_X Z + (Xf)Z) - \nabla_X(f\nabla_Y Z + (Yf)Z) +$$
$$f\nabla_{[X,Y]}Z + ([X,Y]f)Z$$
$$= fR(X,Y)Z + (Y(Xf) - X(Yf) + [X,Y]f)Z$$
$$= fR(X,Y)Z.$$

Thus R is indeed a tensor.

In order to compute the components of R we choose a basis (e_a) and setting $X = e_c$, $Y = e_d$, $Z = e_b$ in (1.9.3) we obtain

$$R(e_c,e_d)e_b = \nabla_d(\nabla_c e_b) - \nabla_c(\nabla_d e_b) + \nabla_{[e_c,e_d]}e_b$$
$$= \nabla_d(\Gamma^f{}_{bc}e_f) - \nabla_c(\Gamma^f{}_{bd}e_f) + \gamma^f{}_{cd}\nabla_f e_b$$
$$= e_d(\Gamma^a{}_{bc})e_a - e_c(\Gamma^a{}_{bd})e_a + \Gamma^f{}_{bc}\Gamma^a{}_{fd}e_a -$$
$$\Gamma^f{}_{bd}\Gamma^a{}_{fc}e_a + \gamma^f{}_{cd}\Gamma^a{}_{bf}e_a.$$

We define the components of the curvature tensor by setting

$$R(e_c,e_d)e_b = R^a{}_{bcd}e_a,$$

which implies that

$$R^a{}_{bcd} = e_d(\Gamma^a{}_{bc}) - e_c(\Gamma^a{}_{bd}) + \Gamma^f{}_{bc}\Gamma^a{}_{fd} - \Gamma^f{}_{bd}\Gamma^a{}_{fc} + \gamma^f{}_{cd}\Gamma^a{}_{bf}.$$
$$(1.9.4)$$

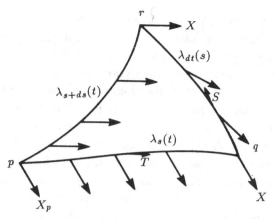

Fig. 1.9.1   On parallel transport around the closed curve pqr the vector X_p changes its value.

In many cases the basis will be coordinate induced, so that the commutator coefficients vanish, and we then obtain the standard formula

$$R^i{}_{kmn} = 2\Gamma^i{}_{k[m,n]} - 2\Gamma^i{}_{p[m}\Gamma^p{}_{|k|n]}, \qquad (1.9.5)$$

where the vertical lines around the suffix k mean that it should be exempted from the antisymmetrization operation.

(1.9.4) Exercise. Let X be a vector field with components X^i with respect to a coordinate induced basis $\partial/\partial x^i$. The components of ∇X have already been defined as $X^i{}_{;k}$. Let the components of $\nabla\nabla X$ be $X^i{}_{;km} = (X^i{}_{;k})_{;m}$. Show that

$$X^i{}_{;km} - X^i{}_{;mk} = R^i{}_{nkm}X^n, \qquad (1.9.6)$$

the *Ricci identity*.

The Ricci identity has an important geometrical interpretation. Let $\lambda_s(t)$ be a 1-parameter family of curves passing through the point $p, (t = 0)$ in M. As s varies the points of constant t define a curve in M denoted by $\lambda_t(s)$. For $t = 0$ $\lambda_t(s)$ degenerates to the point p. Let T, S be the tangent vector fields to the curves $\lambda_s(t), \lambda_t(s)$ respectively. From the geometrical interpretation of the commutator given in section 1.6 it follows that $[T, S] = 0$. Now let X_p be any vector in $T_p(M)$. We may define a vector field X at points on the curves as follows: for any point (s,t) propagate X parallelly along $\lambda_s(t)$ from p, i.e., $\nabla_T X = 0, X(p) = X_p$. Now let $C(ds, dt)$ be the closed curve which starts at p, moves out dt along $\lambda_s(t)$ to $q = (s, dt)$, and then ds along $\lambda_{dt}(s)$ to $r = (s + ds, dt)$, and finally back to p along $\lambda_{s+ds}(t)$. Let $X_p(ds, dt)$ be the vector in $T_p(M)$ obtained from X_p by parallelly transporting it around the closed curve $C(ds, dt)$. The difference $X_p(ds, dt) - X_p$ is a well-defined vector at p which can be computed for infinitesimally small ds, dt as follows. The paths pq and rp do not contribute to this difference because X was parallelly propagated along these curves. Thus the difference is (to first order)

$$X_p(ds, dt) - X_p = X_r - X_q = ds \nabla_S X.$$

(The best way to understand this equation is to write it out explicitly in terms of coordinates as in the geometrical interpetation of the commutator.) Further since $S = 0$ at p, $\nabla_S X = 0$ there. For infinitesimal dt it follows that along qr, $\nabla_S X = dt \nabla_T \nabla_S X$. Now since $[T, S] = 0$ and $\nabla_T X = 0$

$$\nabla_T \nabla_S X = R(S, T)X.$$

Therefore

$$\lim_{\substack{ds \to 0 \\ dt \to 0}} \left[\frac{X_p(ds, dt) - X_p}{ds\,dt} \right] = (R(S, T)X)_p.$$

Thus $R(S, T)X$ measures the change in X after parallel transport around a closed curve. This is intimately related to the fact that if $R \neq 0$ covariant derivatives do not commute. See also the following theorem.

(1.9.5) THEOREM

In a simply connected region U, the curvature R and torsion T vanish if and only if there exists a coordinate system with respect to which the connection coefficients vanish throughout U.

Proof: Suppose that there exists a chart such that $\Gamma^i{}_{km} = 0$ throughout U. Clearly the components of R, T vanish in this chart. Since they are tensor fields they therefore vanish in U. Conversely suppose T, R vanish in a simply connected neighbourhood U of a point p. Let (e_a) be a basis for $T_p(M)$. Since $R = 0$ any element of $T_p(M)$ remains invariant under parallel transport around a closed path in U. Thus if q is any point in U, parallel transport from p to q is path-independent. Therefore the (e_a) generate a basis at each point of U by parallel transport. Clearly $\nabla_a e_b = 0$. If (ω^b) is the dual basis of covectors it follows that $\nabla_a \omega^b = 0$. Now let y^i be an arbitrary coordinate system, and set $\omega^b = e^b{}_i dy^i$. Then

$$0 = \nabla_a \omega^b \Rightarrow e^b{}_{i;k} = 0.$$

But since the torsion tensor vanishes Γ is symmetric in its lower indices. Therefore

$$0 = e^b{}_{[i;k]} = e^b{}_{[i,k]} - \Gamma^n{}_{[ik]} e^b{}_n = e^b{}_{[i,k]}.$$

We may deduce that in a simply connected region $e^b{}_i$ is the gradient of a function, which we may call x^b, so that $\omega^b = dx^b$. It follows that $e_b = \partial/\partial x^b$ and so the original bases were coordinate-induced. But by construction the connection coefficients vanished in this basis.

If the torsion vanishes there is another geometrical interpretation of R which will later prove to be of great significance. First consider two coplanar geodesics in R^n, i.e., straight lines with affine parameter t. Let Z_t be the line joining points of the same t. As t varies the direction and length of Z_t will change linearly. Even if the geodesics are not coplanar Z_t has constant "velocity" and zero "acceleration". This idea carries over directly to a manifold with a connection.

(1.9.6) DEFINITION

*Let X be the tangent vector to a congruence of curves. Any vector Z such that $[X, Z] = 0$ is called a **connecting vector of***

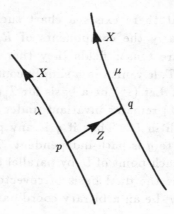

Fig. 1.9.2 $\lambda(t)$ and $\mu(t)$ are neighbouring geodesics of a congruence with tangent vector field X. pq is an infinitesimal curve joining points at the same affine parameter. The "connecting vector" Z is tangent to pq

the congruence. *(The reason for the name comes from the geometrical interpretation of the commutator.)*

(1.9.7) THEOREM

Suppose the torsion T vanishes. Let X be the tangent vector to a congruence of geodesics, and let Z be a connecting vector for the congruence. Then the "acceleration" of Z, $\nabla_X \nabla_X Z$ satisfies the **geodesic deviation equation**

$$\nabla_X \nabla_X Z = -R(X, Z)X. \qquad (1.9.7)$$

Remark. The significance of the name is as follows. Suppose λ, μ are two neighbouring curves of the congruence. Then there is a small number ϵ such that $exp(\epsilon Z)$ maps points of λ to points of μ at the same affine parameter. (1.9.7) then determines the "deviation" of the two geodesics.

Proof: Since $[X, Z] = 0$ and T vanishes, $\nabla_X Z = \nabla_Z X$. Therefore

$$\nabla_X \nabla_X Z = \nabla_X \nabla_Z X$$
$$= \nabla_Z \nabla_X X + R(Z, X)X$$
$$= -R(X, Z)X,$$

since X is geodesic.

The component form of (1.9.7) is

$$\frac{d^2 Z^a}{dt^2} = -R^a{}_{bcd} X^b X^c Z^d, \qquad (1.9.8)$$

where t is an affine parameter for the geodesics. Notice that only $R^a{}_{(bc)d}$ can be determined from experiments based on measurement of geodesic deviation.

Now it is obvious from the definition that $R^a{}_{b(cd)} = 0$. The Riemann tensor has further symmetries.

(1.9.8) THEOREM

Suppose the torsion vanishes. Then

$$R^a{}_{[bcd]} = 0, \qquad (1.9.9)$$

and

$$R^a{}_{bcd} = \tfrac{2}{3}\left(R^a{}_{(bc)d} - R^a{}_{(bd)c} \right). \qquad (1.9.10)$$

*Further the **Bianchi identities***

$$R^i{}_{k[mn;p]} = 0, \qquad (1.9.11)$$

hold.

Proof: Since each of these is a tensor equation it suffices to prove it in any one coordinate system, e.g., normal coordinates. We shall prove (1.9.11), leaving the other two as an exercise for the reader. In normal coordinates we may arrange that $\Gamma^i{}_{(mn)} = 0$

at a point p. Since the torsion vanishes $\Gamma^i{}_{[mn]} = 0$. Thus at p, $\Gamma^i{}_{mn} = 0$. It follows that at p

$$R^i{}_{kmn;p} = \Gamma^i{}_{mk,np} - \Gamma^i{}_{nk,mp},$$

and antisymmetrization now gives the result.

The result (1.9.10) shows that when the torsion vanishes the Riemann tensor can be determined completely from measurements of geodesic deviation. Another tensor that will be required in the sequel is:

(1.9.9) DEFINITION

*The **Ricci curvature tensor** is a $(0,2)$ tensor field obtained by contracting the Riemann tensor*

$$R_{bd} = R^a{}_{bad}. \tag{1.9.12}$$

(1.9.10) Exercise. Show that the Riemann and Ricci tensors for the Newtonian connection (1.8.4) are

$$R^\alpha{}_{0\beta0} = -\phi_{,\alpha\beta}, \qquad R^i{}_{kmn} = 0 \qquad \text{otherwise,} \tag{1.9.13}$$

and

$$R_{00} = -\nabla^2\phi, \qquad R_{ik} = 0 \qquad \text{otherwise,} \tag{1.9.14}$$

where $\alpha, \beta = 1, 2, 3$.

1.10   The pseudo-riemannian metric

The final structure to be imposed on the manifold is a metric. It will be shown that metric structure induces a preferred connection.

(1.10.1) DEFINITION

*A **metric tensor** g at a point p in M is a symmetric $(0,2)$ tensor. It assigns a **magnitude** $\sqrt{|g(X,X)|}$ to each vector X in*

$T_p(M)$, *denoted by* $d(X)$, *and defines the* **angle** *between any two vectors* X, Y *of non-zero magnitude in* $T_p(M)$ *via*

$$a(X,Y) = \arccos \left[\frac{g(X,Y)}{d(X)d(Y)} \right].$$

If $g(X,Y) = 0$ *then* X *and* Y *are* **orthogonal**. *The* **length** *of a curve with tangent vector* X *between* t_1 *and* t_2 *is* $L(t_1, t_2) = \int_{t_1}^{t_2} d(X)dt$. *If* (e_a) *is a basis of* $T_p(M)$, *the components of* g *with respect to the basis are* $g_{ab} = g(e_a, e_b)$.

The simplest example is R^3 with Cartesian coordinates (x, y, z), where

$$g_{ij} = \text{diag}(1, 1, 1),$$

has all the properties. In the Minkowski space of special relativity with (t, x, y, z) coordinates then

$$\eta_{ik} = \text{diag}(1, -1, -1; -1),$$

has the right properties.

If $g(X,Y) = 0$ for all Y in $T_p(M)$ implies that $X = 0$, then g is said to be **non-degenerate**. Henceforth this will be implicitly assumed to be the case. It follows that there is a unique $(2,0)$ tensor g with components g^{ab} satisfying

$$g^{ab}g_{bc} = \delta^a{}_c,$$

i.e., the matrix (g^{ab}) is the inverse of the matrix (g_{ab}). Since both are non-degenerate we have a (non-natural) isomorphism, $T_p(M) \leftrightarrow T_p^*(M)$, $\eta \leftrightarrow X$, given by

$$X^a = g^{ab}\eta_b, \qquad \eta_a = g_{ab}X^b. \qquad (1.10.1)$$

We therefore identify the covector and vector, calling both X. A similar isomorphism holds for tensors, e.g.

$$T^{ab} \leftrightarrow T^a{}_b \leftrightarrow T_a{}^b \leftrightarrow T_{ab},$$

where the indices are raised and lowered with g^{ab}, g_{ab} respectively. (NB: the left-right ordering is preserved!) In general such associated tensors will be regarded as representations of the same object.

(1.10.2) DEFINITION

*The **signature** of a metric is the difference between the number of positive and number of negative eigenvalues. If in a n-dimensional manifold the signature is n then g is **positive-definite** or **Riemannian**, while if the signature is $\pm(n-2)$ then g is **Lorentzian** or **pseudo-riemannian**. A standard result from linear algebra, "Sylvester's law of inertia", guarantees that the signature is basis-independent.*

Accepting for the moment that special relativity has local validity, we deduce that spacetime has dimension 4 and Lorentzian signature. Using standard results from linear algebra we see that at any point p coordinates can be chosen so that the metric takes the form

$$g_{ik} = \mathrm{diag}(1, -1, -1, -1),$$

there. It is now no longer true that $g(X,X) = 0$ implies that $X = 0$, and just as in special relativity we classify vectors at a point as **timelike**, **null** or **spacelike** according as $g(X,X) > 0$, $g(X,X) = 0$, $g(X,X) < 0$.

(1.10.3) Exercise. Show that in Minkowski spacetime the null vectors form a double cone, the *light cone* separating the timelike from the spacelike vectors. Deduce that in a curved manifold of dimension 4 with lorentzian signature, a ***Lorentzian manifold***, the same structure exists in the tangent space of each point.

(1.10.4) Exercise. X, Y are orthogonal vectors in a Lorentzian manifold. Show that if X is timelike then Y is spacelike, whereas if X is null, Y is spacelike or null, but if X is spacelike then Y can be timelike, null or spacelike.

The existence of a metric means that there is a preferred connection with zero torsion. This is constructed explicitly in the following theorem.

(1.10.5) THEOREM

*If a manifold possesses a metric g then there is a unique symmetric connection, the **Levi-Civita connection** or **metric connection** ∇ such that*

$$\nabla g = 0. \tag{1.10.2}$$

Proof: We shall demonstrate existence and uniqueness via an explicit construction. Suppose ∇ is metric. Let X, Y, Z be vector fields. Then since $g(Y, Z)$ is a function

$$
\begin{aligned}
X\left(g(Y, Z)\right) &= \nabla_X\left(g(Y, Z)\right) \\
&= (\nabla_X g)(Y, Z) + g(\nabla_X Y, Z) + g(Y, \nabla_X Z) \\
&= g(\nabla_X Y, Z) + g(Y, \nabla_X Z).
\end{aligned}
$$

Similarly

$$Y\left(g(Z, X)\right) = g(\nabla_Y Z, X) + g(Z, \nabla_Y X),$$

$$Z\left(g(X, Y)\right) = g(\nabla_Z X, Y) + g(X, \nabla_Z Y).$$

Adding the first two equations and subtracting the third gives

$$
\begin{aligned}
g(Z, \nabla_X Y) = \tfrac{1}{2}(&-Z(g(X, Y)) + Y(g(Z, X)) + X(g(Y, Z)) + \\
&g(Z, [X, Y]) + g(Y, [Z, X]) - g(X, [Y, Z])),
\end{aligned}
$$

where the symmetry of the connection has been used to set

$$\nabla_X Y - \nabla_Y X = [X, Y] \quad \text{etc.}$$

Now if (e_a) is a vector basis we may set $Z = e_a$, $X = e_c$, $Y = e_b$ finding

$$
\begin{aligned}
\Gamma_{abc} = g(e_a, \nabla_c e_b) &= g_{ad}\Gamma^d{}_{bc} \\
&= \tfrac{1}{2}(e_b(g_{ac}) + e_c(g_{ba}) - e_a(g_{cb}) + \\
&\quad \gamma^d{}_{cb}g_{ad} + \gamma^d{}_{ac}g_{bd} - \gamma^d{}_{ba}g_{cd}). \tag{1.10.3}
\end{aligned}
$$

If the basis is coordinate induced the last three terms vanish and we have the famous formula defining **Christoffel symbols**

$$\Gamma^i{}_{km} = \tfrac{1}{2}g^{in}(g_{mn,k} + g_{kn,m} - g_{km,n}). \tag{1.10.4}$$

Whenever M possesses a metric we shall usually use the metric connection without explicitly saying so. However even in a metric manifold not all connections are metric, as we shall soon see!

If M is metric the Riemann tensor associated with the metric connection has further symmetries. We define

$$R_{abcd} = g_{ae}R^e{}_{bcd}.$$

(1.10.6) THEOREM

The Riemann tensor associated with a metric connection has the additional symmetries

$$R_{(ab)cd} = 0, \qquad R_{abcd} = R_{cdab}. \tag{1.10.5}$$

Proof: Since these are tensor equations it suffices to verify them for a normal coordinate system for which

$$R_{ikmn} = \tfrac{1}{2}(g_{im,kn} + g_{nk,im} - g_{mk,in} - g_{in,km}). \tag{1.10.6}$$

Equations (1.10.5) then follow immediately.

There are two more curvature quantities that can be defined in a manifold with a metric:

(1.10.7) DEFINITION

*The **Ricci scalar** is*
$$R = g^{ab}R_{ab}. \tag{1.10.7}$$

*The **Einstein curvature tensor** is*

$$G_{ab} = R_{ab} - \tfrac{1}{2}Rg_{ab}. \tag{1.10.8}$$

(1.10.8) Exercise. Show that the contraction of the Bianchi identities (1.9.11), the **contracted Bianchi identities** are

$$0 = \nabla_a G^a{}_b = \nabla_a R^a{}_b - \tfrac{1}{2} e_b(R). \qquad (1.10.9)$$

(1.10.9) Exercise. Propose a definition of a metric for Newtonian spacetimes. Is the Newtonian connection defined by (1.8.4) a metric connection?

An *isometry* is a symmetry of the metric tensor i.e., g is invariant under displacements along the integral curves of some smooth vector field K or, in the language of section 1.6, K preserves g

$$\mathcal{L}_K g_{ab} = 0. \qquad (1.10.10)$$

(1.10.10) Exercise. Show that (1.10.10) implies

$$\nabla_{(a} K_{b)} = 0. \qquad (1.10.11)$$

The property (1.10.11) is **Killing's equation** and its solutions are called **Killing (co-)vectors.**

Isometries are intimately related to conserved quantities. This is a deep result usually referred to as Noether's theorem. Various statements of the theorem of greater or less generality can be found in the literature, e.g., (Arnol'd, 1978). We now verify this for two particular cases. Consider first a freely falling particle whose geodesic worldline has tangent vector P. Define $E = P^a K_a$, where K is a Killing covector. Then

$$
\begin{aligned}
\nabla_P E &= P^a \nabla_a (P^c K_c) \\
&= P^a P^c \nabla_a K_c && \text{since } P \text{ is geodesic,} \\
&= P^a P^c \nabla_{(a} K_{c)} && \text{by symmetry,} \\
&= 0 && \text{since } K \text{ is Killing.}
\end{aligned}
$$

Thus E is conserved along the worldlines of P. As a second example suppose that T^{ac} represents the (symmetric) energy-momentum tensor of a continuous distribution of matter, satisfying $\nabla_c T^{ac} = 0$. If K is a Killing covector define the associated current as $J^a = T^{ac} K_c$. Then by an almost identical calculation one may verify that $\nabla_a J^a = 0$, i.e., the current is conserved.

A very common isometry is when g is invariant in time, i.e., there exists a time coordinate t, and a vector $T = \partial/\partial t$ such that $\pounds_T g = 0$. Such spacetimes are called **static**. Another common example is axisymmetry. If ϕ is an azimuthal angle about some axis, and $\pounds_A g = 0$, where $A = \partial/\partial\phi$, then the spacetime is **axisymmetric**.

1.10.11 Exercise. Identify the conserved quantities E and J^a for static and for axisymmetric spacetimes.

Killing covectors K satisfy a second order equation

$$\nabla_a \nabla_b K_c = R^d{}_{acb} K_d. \qquad (1.10.12)$$

This can be verified by using the identity $R^a{}_{[bcd]} = 0$ and normal coordinates.

1.10.12 Exercise. Consider Minkowski spacetime with Cartesian coordinates x^i. Use (1.10.12) to show that the components of any Killing covector are linear functions of the x^i. (This result was used in exercise 1.6.13 to derive the Poincaré invariance group.)

1.11   Newtonian spacetimes

From a classical point of view, spacetime, the arena in which material processes are imagined to take place, is a 4-dimensional smooth real manifold M. **Absolute time** is a smooth map $t : M \to R$ with non-vanishing gradient, $dt \neq 0$. t is unique up to linear transformations $t' = at + b$ with $a > 0$. Two points or **events** p, q are **simultaneous** if and only if $t(p) = t(q)$. The set of points simultaneous with p forms a **space section** S_p dividing M into two regions, the **past** and **future** of p. In each S_p a Euclidean 3-metric h with signature $(+ + +)$ is defined. There is a lot of experimental evidence that the space sections are flat, e.g., Euclidean geometry. We shall usually choose a coordinate system x^α such that $h_{\alpha\beta} = \mathrm{diag}(1, 1, 1)$, i.e., indices can be raised

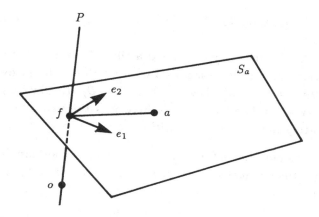

Fig. 1.11.1   The coordinate system in a Galilean spacetime.

and lowered at will in the space sections. The structure (M, t, h) is a **Galilean manifold.**

Let A be a vector field. In general $A(t) \neq 0$ and we say that A is **future pointing, spacelike** or **past pointing** according as $A(t) > 0, = 0, < 0$. Spacelike vectors are then the "vectors" of classical physics. Timelike paths (integral curves of everywhere non-spacelike vectors) represent the **history** or motion of **particles. Galilean coordinates** are defined as follows. First choose a smooth timelike path P representing the history of the origin and a set of 3 spacelike vectors (e_α) at each point of P, and a particular point o on P. Let a be an arbitrary event. The surface S_a intersects P in a unique point f. $\overrightarrow{fa}$ is a spacelike vector and so there exist 3 numbers x^α such that $\overrightarrow{fa} = x^\alpha e_\alpha$. We adjust proper time so that $t(o) = 0$. Let $t = t(f)$. Then the coordinates of a are $x^b(a) = (t, x^\alpha) = (t, x^1, x^2, x^3)$. (P, o, e_α) forms a **Galilean reference frame.** We must investigate what happens to the coordinates when a different frame is chosen. Two frames F, F' have their coordinate systems related by

$$t' = t + \text{const.}, \qquad x'^\alpha = O^\alpha{}_\beta(t)x^\beta + d^\alpha(t), \qquad (1.11.1)$$

where O is a time-dependent orthogonal matrix representing rotations of (e_α) and $d^\alpha(t)$ represents the change in f, i.e., translations.

The transformations (1.11.1) form a group, the **kinematical group** .

In order to do mechanics we need a time derivative for vectors. The covariant derivative associated with h will not do, (why?) and the remaining alternative is $\mathcal{L}_T$, where $T = \partial/\partial t$. Thus a particle has velocity $\dot{x}^a = \mathcal{L}_T x^a = (1, \mathbf{v})$, where $\mathbf{v}$ is the usual "velocity". Unfortunately however this definition is frame-dependent.

(1.11.1) Problem. Let $d/dt = \mathcal{L}_T = \partial/\partial t$ in a Galilean frame F and let $\mathbf{a}(t)$ be a spacelike vector. If F' is a second Galilean frame show that

$$\frac{d\mathbf{a}}{dt'} = \frac{d\mathbf{a}}{dt} + \omega \times \mathbf{a}, \qquad (1.11.2)$$

where ω is a certain spacelike vector, and interpret this equation. [Hint: first use (1.11.1) to express (dt', dx'^{α}) in terms of (dt, dx^{α}). Then compute the components of T' in the frame F. Finally evaluate the left hand side of (1.11.2) in this frame.]

(1.11.2) Exercise. Let $\mathbf{x}(t)$ describe the path of a particle. Show that $\dot{\mathbf{x}}$ is frame dependent, but if two particle paths 1, 2 intersect at a point p, then $\Delta \dot{\mathbf{x}} = \dot{\mathbf{x}}_1 - \dot{\mathbf{x}}_2$ evaluated at p is frame-independent. Show further that the acceleration $\ddot{\mathbf{x}}(t)$ of a particle is frame-dependent, but for two particle paths **touching** at p, (i.e., with the same velocity,) then $\Delta \ddot{\mathbf{x}}$ is frame-independent.

The main result in classical mechanics is **Newton's law of motion** : there exist special frames **inertial frames** such that the acceleration of a particle of mass m is related to a force vector $\mathbf{f}$ by

$$m\ddot{\mathbf{x}} = \mathbf{f}. \qquad (1.11.3)$$

(1.11.3) Exercise. Under what subset of the group of transformations (1.11.1) is equation (1.11.3) invariant?

In order to make (1.11.3) frame-independent we have, following exercise 1.11.2, to introduce a universal standard acceleration field $\mathbf{a}$, depending on position and velocity, and write equation (1.11.3) as

$$m\left(\ddot{\mathbf{x}} - \mathbf{a}(\mathbf{x}, \dot{\mathbf{x}})\right) = \mathbf{f}, \qquad (1.11.4)$$

which is now frame-independent. Inertial frames are those in which **a** vanishes.

How can one measure **a**? Consider laboratory mechanics. Since we cannot shield off gravity we have, in any inertial frame

$$m(\ddot{\mathbf{x}} - \mathbf{a}) = \mathbf{f} + \mu\mathbf{g},$$

where **g** is the gravitational acceleration and μ is the passive gravitational mass of the particle. We rewrite this as

$$\ddot{\mathbf{x}} = \mathbf{a} + \frac{\mu}{m}\mathbf{g} + \frac{1}{m}\mathbf{f}.$$

In general the frame one uses in a laboratory is not an inertial one, and using (1.11.2) twice we have

$$\ddot{\mathbf{x}} = \mathbf{a} + \frac{\mu}{m}\mathbf{g} + \frac{1}{m}\mathbf{f} - 2\omega \times \dot{\mathbf{x}} - \dot{\omega} \times \dot{\mathbf{x}} - \omega \times (\omega \times \mathbf{x}). \quad (1.11.5)$$

By doing enough experiments with varying $\mathbf{x}$, $\dot{\mathbf{x}}$ we can certainly measure $\mathbf{f}/m$, ω, $\mathbf{a} + (\mu/m)\mathbf{g}$. Now by considering particles made from different materials we might hope to measure μ/m, and hence **a**, **g** separately. However μ/m is not measurable! The Eötvös experiment shows that to one part in 10^{12} μ/m is independent of the material, and so by a suitable choice of units, $\mu/m = 1$. Thus only $\mathbf{a} + \mathbf{g}$ can be measured, and not **a**, **g** separately. **a** has no separate physical meaning; we can and must incorporate it into gravity. We now reformulate Newton's law as follows. There exists *free-fall motions* which are characterized with respect to a special class of galilean frames, the ***Newtonian frames*** by the equation of motion

$$\ddot{\mathbf{x}} = -\nabla\phi, \quad (1.11.6)$$

in which the *gravitational potential* ϕ is a frame-dependent smooth function. If a force represented by a frame-independent spacelike vector **f** is present then (1.11.6) becomes

$$m(\ddot{\mathbf{x}} + \nabla\phi) = \mathbf{f}. \quad (1.11.7)$$

(1.11.4) Exercise. What subset of the group of transformations (1.11.1) relates Newtonian frames by preserving (1.11.7)? Show further that under a uniform acceleration **a**, ϕ transforms as

$$\phi' = \phi - \mathbf{x}.\mathbf{a}. \quad (1.11.8)$$

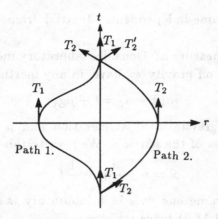

Fig. 1.11.2   Two geodesic paths in a Newtonian spacetime.

We can rephrase these results in geometrical form. We rewrite the free-fall law (1.11.6) as

$$\ddot{x}^i + \Gamma^i{}_{km}\dot{x}^k\dot{x}^m = 0,$$

where

$$\Gamma^\alpha_{00} = \phi_{,\alpha}, \qquad \Gamma^i{}_{km} = 0 \quad \text{otherwise},\qquad (1.11.9)$$

and by convention greek indices range over $1, 2, 3$. This **Newtonian connection** is symmetric and so the torsion vanishes in Newtonian gravity. A simple calculation, exercise 1.9.10, shows that

$$R^\alpha{}_{0\beta0} = -\phi_{,\alpha\beta}, \qquad R^i{}_{kmn} = 0 \quad \text{otherwise}.\qquad (1.11.10)$$

It follows from theorem 1.10.6 that there is no metric of which R us the Rieman tensor. A physical interpretation of R has been given by Ehlers, 1973. Consider two particles at the north pole. Suppose there is a smooth straight tube leading to the south pole, and particle 1 is dropped from rest down it. It just reaches the south pole and then returns to the north pole. Particle 2 is thrown vertically upwards with sufficient velocity so as to return to the north pole simultaneously with particle 1. Consider the spacetime diagram for this experiment, figure 1.11.2. Transporting the initial tangent vector T_1 to path 1 parallelly along path 1 always produces a vector tangent to path 1, since path 1 is a geodesic. The same holds for path 2. If however T_2 is transported along path 1 we obtain the vector T_2' at the end of the experiment. To see that $T_2' \neq T_2$, note first that a parallelly propagated basis remains a basis. Initially $T_2^r > 0$ and finally $T_2^r < 0$. Thus if $T_2' = T_2$ then there exists a point q on path 1 at which the parallelly transported T_2 satisfies $T_2^r = 0$. Therefore at q T_1, T_2 do not form a basis for vectors in the $t - r$ subspace, whereas they do initially and finally. Thus parallel transport of timelike vectors is path-dependent. However using (1.11.9) and the geometrical interpretation of the Riemann tensor given in section 1.9 it is easy to see that parallel transport of spacelike vectors is path-independent.

By moving to a suitably accelerated frame we can always arrange, using (1.11.8), that $\phi_{,\alpha} = 0$ so that the connection vanishes for this frame. Thus normal coordinates correspond to freely falling frames. Next consider two points at x^α, $x^\alpha + \Delta x^\alpha$. The relative gravitational acceleration between them is

$$\Delta \ddot{x}^\alpha = \phi_{,\alpha\beta} \Delta x^\beta = -R^\alpha{}_{0\beta0} \Delta x^\beta,$$

or

$$\Delta \ddot{x}^i = -R^i{}_{0k0} \Delta x^k. \tag{1.11.11}$$

Since this is a tensor equation the relative gravitational acceleration or tidal force is a tensor which cannot be transformed away. Indeed (1.11.11) is the equation of geodesic deviation.

Finally consider the relation of the gravitational field to the sources producing it. There is a lot of experimental evidence at a terrestrial and solar system level for **Poisson's law**

$$\nabla^2 \phi = 4\pi G \rho. \tag{1.11.12}$$

Using (1.11.9) this becomes

$$R_{00} = -4\pi G \rho, \tag{1.11.13}$$

an equation to which we shall return.

1.12   Special relativity

The postulates of special relativity are that spacetime is a smooth
4-manifold M with a Lorentzian flat metric η_{ab} such that in certain
coordinate systems, *inertial coordinates*

$$\eta_{ab} = \text{diag}(1, -1, -1, -1).$$

The null cones of η describe light propagation in vacuum, timelike
geodesics describe free fall motions and

$$\int d\tau = \int \sqrt{\eta_{ab}\, dx^a dx^b} = \int \sqrt{1 - v^2}\, dt$$

taken along a timelike path representing a particle gives the
proper time as measured by a standard clock associated with
that particle.   Inertial coordinates are defined up to linear
transformations

$$x'^a = L^a{}_b x^b + d^a, \tag{1.12.1}$$

where $d^a, L^a{}_b$ are constants and $L^T \eta L = \eta$. These transformations
form a group, the **Poincaré group**. Simultaneity and causality
are determined by the null cones. Special relativity is well verified
by **local** experiments in high energy physics.

Under conditions where gravitational effects can be ignored
there is also extremely good evidence that Maxwell's equations
are valid. Let (x^i) be an inertial coordinate system. Let the components of the electromagnetic field be E^α, B^α, the electric charge
density be ρ, and the current density be j^α. These quantities can
be combined into a Maxwell field tensor $F_{ik} = F_{[ik]}$ and a 4-current
density J^i according to

$$E_\alpha = F_{0\alpha}, \qquad B_\alpha = -\tfrac{1}{2}\epsilon_{\alpha\beta\gamma}F^{\beta\gamma}, \qquad J^0 = \rho, \qquad J^\alpha = j^\alpha,$$

and Maxwell's equations take the form

$$F_{[ik,m]} = 0, \qquad F^{ik}{}_{,k} = 4\pi J^i. \tag{1.12.2}$$

The **energy density** of the electromagnetic field is $W = \tfrac{1}{2}(E^2 + B^2)$, the **Poynting vector** is $\mathbf{S} = (1/4\pi)\,\mathbf{E} \times \mathbf{B}$ and the
Maxwell stress tensor is

$$t_{\alpha\beta} = \left(\frac{1}{4\pi}\right)\left(\tfrac{1}{2}(E^2 + B^2)\delta_{\alpha\beta} - E_\alpha E_\beta - B_\alpha B_\beta\right).$$

In a source-free region these quantities satisfy

$$\dot{W} + \nabla . \mathbf{S} = 0, \qquad \dot{\mathbf{S}} + \nabla . t = 0,$$

corresponding to conservation of energy and momentum. The *energy-momentum tensor* is defined by

$$T^{ik} = \left(\frac{1}{4\pi}\right)\left(\tfrac{1}{4}\eta^{ik}F_{mn}F^{mn} - F^{im}F^{kn}\eta_{mn}\right). \qquad (1.12.3)$$

In an inertial coordinate system $T^{00} = W$, $T^{\alpha 0} = S^{\alpha}$ and $T^{\alpha\beta} = t^{\alpha\beta}$. The conservation laws become

$$T^{ik}{}_{,k} = 0. \qquad (1.12.4)$$

Next we shall consider fluids in special relativity. The energy density is the sum of the rest mass density and the thermal energy density. Under normal laboratory conditions the second contribution is negligible in comparison with the first, but it cannot be ignored in astrophysics and cosmology. We generalize the idea of fluid velocity, replacing $\mathbf{u}$ by U^i, a unit timelike vector, $U^iU_i = 1$. We start by considering the *projection tensor*

$$h^{ik} = \eta^{ik} - U^iU^k. \qquad (1.12.5)$$

(1.12.1) Exercise. Show from the definition that

i) $h^{[ik]} = 0$,
ii) $h^i{}_i = 3$,
iii) $h^{ik}U_k = 0$,
iv) $h^i{}_m h^m{}_k = h^i{}_k$.

To understand the name let $p \in M$ and consider h^i_k as a map $T_p \to T_p$. Let N be the space of all vectors proportional to U and let W be the space of all vectors orthogonal to U. For any vector V

$$V^i = (U^kV_k)U^i + h^i{}_kV^k,$$

where the first term is in N, the second in W. Thus h is a map $T_p \to W$. The *energy-momentum tensor of a perfect fluid* is taken to be

$$T^{ik} = \rho U^iU^k - ph^{ik}. \qquad (1.12.6)$$

The conservation equations are presumably $T^{ik}{}_{,k} = 0$. To verify this consider

$$0 = T^{ik}{}_{,k}$$
$$= (\rho U^k)_{,k} U^i + \rho U^i{}_{,k} U^k - p^{,i} + U^i p_{,k} U^k + p(U^i U^k{}_{,k} + U^i{}_{,k} U^k).$$

Multiplying by U_i gives

$$0 = -(\rho U^k)_{,k} - p^{,i} U_i + p_{,k} U^k + p U^k{}_{,k},$$

or

$$\rho_{,k} U^k + (\rho + p) U^k{}_{,k} = 0,$$

or

$$\dot{\rho} + (\rho + p)\theta = 0, \qquad\qquad (1.12.7)$$

where $\dot{\rho}$ denotes $\rho_{,i} U^i$, the convective derivative along U, and $\theta = U^i{}_{,i}$. An alternative form for θ is $d \ln \delta V / dt$ where δV is a small comoving volume element. Thus equation (1.12.7) can be written

$$\dot{\rho}\delta V + (\rho + p)(\delta V)^{\cdot} = 0,$$

or

$$(\delta E)^{\cdot} + p(\delta V)^{\cdot} = 0,$$

where $\delta E = \rho \delta V$ is the energy contained in the volume element. This is the first law of thermodynamics. Next consider

$$h^{im} T^k{}_{m,k} = 0,$$

or

$$(\rho + p)\dot{U}^i - h^{ik} p_{,k} = 0, \qquad \dot{U}^i = U^i{}_{,k} U^k,$$

the Euler equation, provided that $\rho + p$ is interpreted as the inertial mass density.

Thus it may be concluded that in special relativity a $(2,0)$ tensor T^{ik}, the **energy-momentum tensor** can be associated with any system of particles, or fluid, or any other kind of field, (except gravity). T^{ik} satisfies

$$T^{ik} = T^{(ik)}, \qquad T^{ik}{}_{,k} = 0, \qquad T_{ik} V^i V^k > 0 \quad \text{for all timelike } V^k.$$
$$(1.12.8)$$

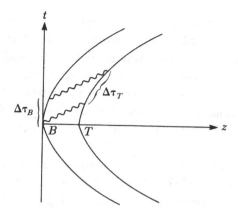

Fig. 1.13.1   The Pound-Rebka experiment in case b). The worldlines of the base B and top T of the tower are hyperbolae. If two photons are emitted upwards from the base $\Delta\tau_B$ apart, they are received with a larger time separation $\Delta\tau_T$ corresponding to a red shift.

1.13   The general theory of relativity

It has been concluded that Newtonian theory is inadequate to describe high velocity phenomena which require special relativity while special relativity disregards gravitation. Further while we can define a (relativistic) inertial frame at each point, the relative orientation of these frames is not well-determined. It is clear that these two weaknesses are related to each other. Gravitation modifies Newtonian theory by preventing exact inertial frames from existing and permits only approximate local inertial frames attached to freely falling particles, and these local inertial frames are accelerated relative to each other. This suggests that the incorporation of gravity into relativity theory requires a change of geometry such that the system of timelike straight lines is distorted into a system of curved lines representing relatively accelerated freely falling particles. There are two important experiments which support this viewpoint.

It is observed that light rays passing close to the sun are deflected and bent towards the sun. Thus the actual light cones in a gravitational field are not the null cones of the Minkowski metric. The **Pound-Rebka experiment** shows that the preferred frames of special relativity coincide with the local gravitational inertial frames. In this experiment γ-ray photons were emitted vertically upwards by nuclei situated at the base of a tower, and absorbed by similar nuclei at the top. A change in frequency was observed. Let B, T be the bottom and top of the tower of height l, let P be the photon, F a freely falling body and let U_e, U_a be the 4-velocities of the emitter and absorber respectively, see figure 1.13.1. The analysis of the experiment, assuming special relativity is a sufficiently good approximation, depends on whether the inertial frames of special relativity are a) attached to the earth, or b) attached to freely falling particles, or c) some other possibility. In case a) the worldlines of B, T are parallel straight lines. In case b) if we ignore the slight variation of g with height the worldlines of B, T are hyperbolae with common null asymptotes. In both cases the path of P is a null straight line.

(1.13.1) Exercise. Show, using special relativity, that the frequency shifts (red shifts) in the two cases are a) 0, b) $\delta\lambda/\lambda = gl/c^2$ + smaller terms. Experiments have decided in favour of b) with surprisingly good accuracy. Thus there are no global inertial frames in relativity theory, but local gravitational inertial frames are identical with those in which Maxwell's equations are valid locally. (Notice that there is an assumption underlying exercise 1.13.1, namely that the nuclei represent ideal clocks in spite of their acceleration g. This, the *clock postulate*, seems reasonable because the gravitational force on the nuclei is many orders of magnitude smaller than the nuclear forces dominating the emission and absorption.)

These considerations suggest the following hypothesis known as the *strong principle of equivalence*.

(1.13.2) AXIOM

In the neighbourhood of each event e in spacetime there exist preferred coordinate systems called inertial at e. For each class

*of physical phenomena except gravity, (mechanics, electrodynam-
ics, hydrodynamics, atomic physics etc.) a set of local laws can
be selected which, if expressed in terms of inertial coordinates and
evaluated at e take on some standard form with standard constants
independent of the spacetime location and the gravitational tidal
field. (As a corollary, the local inertial coordinates are realizable
approximately by freely falling non-rotating small laboratories.)*

If this principle is accepted it follows that special relativity is
a **local** approximation, valid only when gravitational tidal forces
are negligible. It further implies that non-gravitational local physical laws, if known in the special relativity approximation, can be
carried over to the general case if properly formulated.

The following "heuristic derivation" of Einstein's field equations is due to Ehlers, 1973.

It is clear that a gravitational theory of spacetimes should have
the following features:

a) it agrees locally with special relativity,
b) it singles out a class of preferred relatively accelerated worldlines representing free-fall,
c) it admits the definition of a tidal field tensor related to the
   source density of gravity by a law similar to and consistent
   with the experimentally verified Poisson's law,
d) it is capable of describing light deflection.

The step taken by Einstein in 1915 to satisfy requirement a)
was to assume the existence of a pseudo-riemannian metric g_{ab}.
Newtonian theory suggested that b), c) could be satisfied by introducing a non-integrable symmetric linear connection representing
the gravitational field. (A connection is **integrable** if parallel
transport is path-independent.) However this forced Einstein to
introduce a non-minkowskian metric. For the strong principle
of equivalence, (a precise form of a)) requires that there exist
coordinates x^i in some neighbourhood of each event e such that

$$g_{ik} = \eta_{ik}, \qquad g_{ik,m} = 0, \qquad \Gamma^i{}_{km} = 0 \text{ at } e. \qquad (1.13.1)$$

Now (1.13.1) implies $g_{ik;m} = 0$ at e. This is a tensor equation true in all coordinate systems. Then theorem 1.10.5 implies

that Γ^i_{km} is the unique Levi-Civita connection associated with g_{ik}, and if $g_{ik} = \eta_{ik}$ globally then condition b) is violated. In view of the success of special relativity many physicists are reluctant to introduce a non-minkowskian metric and even when confronted with the above argument, reluctantly admit a g_{ik} but also carry along the old η_{ik} in their theories. The role of η_{ik} in this **Poincaré-invariant** approach is problematic. Its null cone structure cannot represent the propagation of light, and it follows from the Pound-Rebka experiment that η does not determine nuclear proper times. Indeed it turns out that in most of these approaches not only is η non-measurable, but it is not even uniquely related to measurable quantities, and so it is best discarded.

Thus the first main assumption of general relativity theory is the existence of a smooth manifold M with a pseudo-riemannian metric g of signature $(+ - - -)$. The null geodesics of g represent light rays, timelike geodesics represent the world lines of freely falling particles, and the arc-length along timelike curves, representing the (arbitrary) motion of particles, measures the time shown by a standard clock. This assumption satisfies the conditions a), b). It incorporates the experimentally verified features of Newtonian theory, (non-integrable connection, curvature) and of special relativity, (light cones, Minkowskian inner product of 4-vectors) and discards the ill-founded special structures, (absolute time, integrable connection) of the earlier theories. Moreover the strong principle of equivalence connects the theory with local familiar physics.

In special relativity the energy-momentum tensor T_{ik} satisfies $T^{ik}_{,k} = 0$. According to the strong principle of equivalence the corresponding conservation law in the presence of a gravitational field is

$$T^{ik}_{;k} = 0, \tag{1.13.2}$$

for this is the covariant form that reduces to the special relativity form in inertial coordinates.

Condition c) still has to be satisfied. It was shown in section 1.11 that Poisson's law could be written in the form (1.11.13)

$$R_{00} = -4\pi G\rho.$$

Now for a perfect fluid with 4-velocity U^i the covariant form for the left hand side is $R_{ik}U^iU^k$. The energy density is $T_{ik}U^iU^k$ and using the equivalence of energy and mass this is a suitable candidate for ρ. A second quantity which in a weak field is almost equal to the density is $T = T^i{}_i = \rho - 3p$. (In the solar system $p/\rho \leq 10^{-6}$.) Hence for any value of the arbitrary constant λ

$$(\lambda T_{ik} + (1 - \lambda)g_{ik}T)U^iU^k,$$

corresponds very nearly to the Newtonian density. Thus we set

$$R_{ik}U^iU^k = -4\pi G(\lambda T_{ik} + (1 - \lambda)g_{ik}T)U^iU^k.$$

The easiest way to satisfy this equation for all timelike U^i is to require

$$R_{ik} = -4\pi G(\lambda T_{ik} + (1 - \lambda)g_{ik}T). \qquad (1.13.3)$$

The argument for (1.13.3) is not entirely compelling, and this has led to a number of alternative but considerably more complicated theories of gravity.

Fortunately the parameter λ can be determined. We first take the trace of (1.13.3)

$$R = -4\pi G(4 - 3\lambda)T. \qquad (1.13.4)$$

We next take the divergence of (1.13.3) and use the contracted Bianchi identities $R^{ik}{}_{;k} = \frac{1}{2}R^{,i}$ and the conservation law (1.13.2) to obtain, using (1.13.4)

$$4\pi G(1 - \lambda)T_{,i} = 2\pi G(4 - 3\lambda)T_{,i}.$$

In general $T_{,i} \neq 0$ and so $\lambda = 2$. Using the Einstein tensor $G_{ik} = R_{ik} - \frac{1}{2}Rg_{ik}$, (1.13.3) can be written as the **Einstein (1915) field equations**

$$G_{ik} = -8\pi GT_{ik}. \qquad (1.13.5)$$

Notice that in this derivation condition d) has not been used. It can be shown that Einstein's theory does indeed satisfy condition d), and indeed this condition can be used to eliminate almost all of the alternative theories.

A complete description of solutions of (1.13.5) is not yet possible. The remainder of this monograph is concerned with (often quite intricate) techniques for extracting information about gravity. Readers with little experience of Einstein's theory will find a succinct description of some exact solutions in chapter 5 of Hawking and Ellis, 1973, although they may find it helpful to work through the first two sections of chapter 3 here first. Once the first two chapters of this book have been mastered the comprehensive survey of solutions given in Kramer et al., 1980 should prove helpful.

Given such complexity it makes sense to look for problems in which some simplifying feature obtains. The major problem is nonlinearity and so the major simplication occurs when the problem is close to a known exact solution. Here we present linearized theory close to flat spacetime, i.e., where fields are everywhere weak. We regard this as a theory on Minkowski spacetime, and assume a global Cartesian coordinate chart x^i has been selected. We write the metric tensor as

$$g_{ik} = \eta_{ik} + \epsilon h_{ik}, \tag{1.13.6}$$

where ϵ is "small", and for the rest of this chapter higher order terms will be discarded implicitly. For example we write the inverse metric as

$$g^{ik} = \eta^{ik} - \epsilon h^{ik}, \tag{1.13.7}$$

where indices on "small" quantities are raised and lowered with the η-metric. Much of the calculation is routine, and so is left as exercises.

(1.13.3) Exercise. Show that

$$R_{ikmn} = \tfrac{1}{2}\epsilon \left(h_{im,kn} + h_{nk,im} - h_{mk,in} - h_{in,km} \right). \tag{1.13.8}$$

Define

$$\overline{h}_{ik} = h_{ik} - \tfrac{1}{2}\eta_{ik}h_m{}^m. \tag{1.13.9}$$

Next show that the Einstein field equations (1.13.5) become

$$\overline{h}_{ik,m}{}^m + \eta_{ik}\overline{h}_{mn,}{}^{mn} - \overline{h}_{im,k}{}^m - \overline{h}_{km,i}{}^m = -16\pi G\epsilon^{-1}T_{ik}. \tag{1.13.10}$$

Using problem 1.6.14 we deduce that under a change of gauge

$$h_{ik} \mapsto h_{ik} + \mathcal{L}_\xi \eta_{ik} = h_{ik} + 2\xi_{(i,k)} \qquad (1.13.11)$$

where x^i is an arbitrary vector.

(1.13.4) Exercise. Verify explicitly that the curvature tensors are gauge-invariant. Is this surprising?

Under a gauge change (1.13.11), $\overline{h}_{ik,}{}^k \mapsto h_{ik,}{}^k + \Box\xi_i$, where $\Box\xi_i = \xi_{i,m}{}^m$. In flat space $\Box\phi = \rho$ can always be solved and so, working in Cartesian coordinates, we can always choose ξ so as to set $\overline{h}_{ik,}{}^k = 0$. Because of the above exercise we know that the (linearized) curvature tensors will be unaffected. This choice is known as **harmonic gauge** or **de Donder gauge**.

(1.13.5) Exercise. Show that in de Donder gauge the field equations (1.13.10) become

$$\Box \overline{h}_{ik} = -16\pi G\epsilon^{-1} T_{ik}. \qquad (1.13.12)$$

By the above remark this equation can always be solved.

(1.13.6) Exercise. Consider the external gravitational field of a static spherically symmetric body of mass ϵM situated at the coordinate origin $x = y = z = 0$ i.e., its 4-velocity is $U^i = (1,0,0,0)$. Take $T_{ik} = \epsilon M \delta(\mathbf{x}) U_i U_k$. Assuming that the field is both weak and a function of $R = \sqrt{x^2 + y^2 + z^2}$ only derive the linearized metric.

(1.13.7) Exercise. The problem in the previous exercise has an exact solution, the **Schwarzschild metric** appropriate to a mass ϵM. Linearize the exact metric and compare it with the solution of (1.13.6).

This completes our survey of the foundations of Einstein's theory of gravitation.

2

SPINORS

2.1 Introduction

Although the treatment of spinors given here is totally independent of quantum mechanics it may help those with an elementary
understanding of that subject if we recall some of the properties of
the Pauli spin operators. The wave function ψ of an electron has
two complex components corresponding in some particular reference frame to 'spin-up' and 'spin-down' with respect to a chosen
axis. Let $\vec{\sigma}$ be a 2×2 matrix operator in the Hilbert space spanned
by the wave function, which is also a vector in ordinary 3-space,
and so can be represented as a set of three components along the
coordinate axes. In other words if $\vec{u} = (u^1, u^2, u^3)$ is a unit vector
the 'spin' in the direction $\vec{u}$ is

$$\vec{u}.\vec{\sigma} = u^1 \sigma_1 + u^2 \sigma_2 + u^3 \sigma_3.$$

In the Hilbert space of ψ the components of $\vec{\sigma}$ can be represented
by the Pauli spin matrices

$$\sigma_1 = \begin{pmatrix} 0 & 1 \\ 1 & 0 \end{pmatrix}, \qquad \sigma_2 = \begin{pmatrix} 0 & -i \\ i & 0 \end{pmatrix}, \qquad \sigma_3 = \begin{pmatrix} 1 & 0 \\ 0 & -1 \end{pmatrix},$$

so that the total spin operator is the matrix

$$\vec{u}.\vec{\sigma} = \begin{pmatrix} u^3 & u^1 - iu^2 \\ u^1 + iu^2 & -u^3 \end{pmatrix}. \tag{2.1.1}$$

The theory is required to be invariant under transformations of
the form

$$\psi' = Q\psi, \qquad \vec{\sigma}' = Q\vec{\sigma}Q^\dagger, \tag{2.1.2}$$

where Q is in $SU(2)$, and † denotes complex conjugate plus transpose. To obtain the relativistic generalization of this one simply
adds a 'timelike' component to $\vec{\sigma}$ viz. $\sigma_0 = I_2$ where I_2 is the 2×2
unit matrix. Now (2.1.1) becomes

$$\vec{u}.\vec{\sigma} = \begin{pmatrix} u^0 + u^3 & u^1 - iu^2 \\ u^1 + iu^2 & u^0 - u^3 \end{pmatrix}. \qquad (2.1.3)$$

Henceforth we restrict Q to be a member of $SL(2,\mathcal{C})$, so as to
preserve the Hermitian nature of (2.1.3).

Now the determinant of this matrix is precisely the (Lorentzian)
norm of the 4-vector u^a. This determinant is obviously invariant under the transformations (2.1.2) and so every element Q of
$SL(2,\mathcal{C})$ defines a Lorentz transformation. Since $SL(2,\mathcal{C})$ is simply connected, Q is continuous with the identity; so therefore is the
Lorentz transformation, and so the latter must be **proper**, i.e., it
preserves both time and space orientation. Further this map from
$SL(2,\mathcal{C})$ to the Lorentz group $L(4)$ is $(2,1)$ for $-Q$ defines the
same Lorentz transformation.

Unlike Dirac we choose not to form 4-component objects from $\vec{\sigma}$.
Instead we regard (2.1.3) as a method for encoding real 4-vectors
into 2×2 Hermitian matrices

$$u^a \mapsto \begin{pmatrix} u^{00'} & u^{10'} \\ u^{01'} & u^{11'} \end{pmatrix}. \qquad (2.1.4)$$

We posit the existence of more fundamental objects, complex 2-
vectors which will be called **spinors**. It will turn out that all real
spacetime vectors and tensors can be described in terms of the
tensor algebra over these 2-vectors, their complex conjugates and
their duals. The Dirac 4-spinor formalism can also be formulated
in terms of these 2-spinors, as will be described in appendix A.
On the other hand spinors need not be describable in terms of
conventional tensors, although they will be in orientable causal
spacetimes of signature $(+ - - -)$. Therefore we shall adopt the
viewpoint that (2-component) spinors are more basic than tensors
or Dirac spinors.

2.2 Spinor algebra

Let S be a 2-dimensional symplectic vector space over $\mathcal{C}$, as defined below.

(2.2.1) DEFINITION

*A **symplectic linear structure** on an even-dimensional vector space S is a non-degenerate bilinear skew-symmetric 2-form in S. This form is called the **skew-scalar product** and is denoted by $[\xi, \eta] = -[\eta, \xi]$. (A 2-form $[\ ,\ ]$ on S is **non-degenerate** if $[\xi, \eta] = 0$ for all η in S implies $\xi = 0$.) The space S with the symplectic structure $[\ ,\ ]$ is called a **symplectic vector space**. A linear transformation $Q : S \rightarrow S$ is called **symplectic** if it preserves the skew-scalar product,*

$$[Q\xi, Q\eta] = [\xi, \eta], \qquad \forall \ \xi, \ \eta.$$

*The set of symplectic transformations forms the **symplectic group** of appropriate dimension.*

As we shall see $Sp(2)$, the symplectic group of dimension 2 is isomorphic to $SL(2, \mathcal{C})$, which provides the link to the previous section. A symplectic structure may be unfamiliar and so we shall outline some of the basic features. With the standard Euclidean scalar product no non-zero vector is self-orthogonal; with a Lorentzian scalar product every null vector is self-orthogonal, but with a skew-scalar product every vector is self-orthogonal. Indeed since S is 2-dimensional the space of vectors orthogonal to a non-zero vector ξ consists precisely of vectors proportional to ξ. For if η were orthogonal to ξ and not proportional to it the two vectors would constitute a basis for S and so every vector would be orthogonal to ξ, which would contradict the non-degeneracy condition.

Let S^* be the dual space of S. We build the tensor algebra over S, S^* in the usual way. Given bases of S and S^* we may define a non-natural isomorphism between the two spaces: two vectors are regarded as being the 'same' if their components with respect to the two bases are identical. However the skew-scalar product defines a **natural** isomorphism: to the element ξ of S we associate $[\xi,\ ]$ in S^* which is a linear map $S \to C : \eta \mapsto [\xi, \eta]$. (A demonstration that it is an isomorphism will be given later.)

Unfortunately the geometrical properties of symplectic spaces are not widely known, although a lucid introduction can be found in section 41 of Arnol'd, 1978. Therefore it is advantageous initially to do calculations with respect to bases. A standard basis may be constructed as follows. Let o be an arbitrary vector in S. Let ι be any non-parallel vector. From the paragraph above it follows that $[o, \iota] \neq 0$ and so we may normalize ι to set $[o, \iota] = 1$. Clearly ι is not unique, for we may add an arbitrary multiple of o to it without upsetting the normalization.

(2.2.2) DEFINITION

*If o is a non-zero vector and ι is chosen to set $[o, \iota] = 1$ then (o, ι) is a **spin basis** for S.*

Given an arbitrary spin basis we may define components of vectors with respect to the basis. If ξ is in S the **components of ξ** are ξ^A, $A = 0, 1$, where

$$\xi = \xi^0 o + \xi^1 \iota. \tag{2.2.1}$$

Clearly

$$o^A = (1, 0), \qquad \iota^A = (0, 1).$$

Pure mathematicians often abhor the use of components, because of their ambiguity; does $u^a + v^a$ represent (a component of) a vector sum, or simply the addition of two components, i.e., numbers? However they have been unable to provide an alternative notation capable of the same expressive powers. (See e.g., the discussion in Schouten, 1954.) We shall use the 'suffix notation' whenever it is the most convenient approach. An obvious example is the natural symplectic isomorphism between S and S^* ; we shall identify isomorphic partners by the the same (greek) kernel

character, while the position of the suffix identifies to which space
the element belongs. Thus ξ^A ($\in S$) is isomorphic to ξ_A ($\in S^*$).

We can get round the mathematicians' difficulty by introducing
Penrose's 'abstract index' notation. Briefly this involves the following. We build an infinite set of copies of S labelled $S^A, S^B, \ldots$
etc. We also need a "sameness" map (implied by the concept of
copies) which identifies two vectors in different spaces. We also
do the same for the dual spaces. Then the notation is as follows:

ξ^A, a vector in S^A,

η^A, another vector in S^A,

$\xi^A + \eta^A$, their sum, also in S^A,

ξ^B, the "same" vector as ξ^A, but in S^B,

η_A, the dual, i.e., the isomorphic partner induced by the
   symplectic structure, in S_A,

$\eta_A\xi^A$, the same as $[\eta, \xi]$ in $\mathcal{C}$.

A complete rigorous account is given in Penrose and Rindler,
1984. Indeed mathematical physicists have been using this notation for tensors without knowing it. Now to satisfy the pure
mathematicians we should rewrite equation (2.2.1) as follows. We
first introduce a spin basis $e_{\hat{A}}{}^A$ where the lower hatted index labels
the 2 members of the basis, and the upper index indicates that
the basis members are in S. A suitable choice is e.g., $e_{\hat{0}}{}^A = o^A$,
$e_{\hat{1}}{}^A = \iota^A$. We then write

$$\xi^A = \xi^0 e_{\hat{0}}{}^A + \xi^1 e_{\hat{1}}{}^A = \xi^{\hat{A}} e_{\hat{A}}{}^A.$$

While rigorous, this notation becomes increasingly more difficult
to read as equations become more complicated, so we shall adopt
a policy of partial vacillation. Henceforth spinorial and tensorial
indices with a caret always refer to a basis. Indices without a caret
can be given either interpretation unless an explicit statement
to the contrary is given. Although this sounds ambiguous we
shall develop the notation in such a way as to prevent actual
ambiguities.

Clearly the symplectic structure can be identified with an el-
ement of $S^* \times S^*$ which we write as $\epsilon_{AB} = -\epsilon_{BA}$, so that

$[\xi, \eta] = \epsilon_{AB} \zeta^A \eta^B$. The condition that (o, ι) is a spin basis for S translates into

$$\epsilon_{AB} o^A o^B = \epsilon_{AB} \iota^A \iota^B = 0, \qquad \epsilon_{AB} o^A \iota^B = 1.$$

In components with respect to this basis we have

$$\epsilon_{AB} = \begin{pmatrix} 0 & 1 \\ -1 & 0 \end{pmatrix}.$$

This is a non-singular matrix, and we identify its inverse, up to a conventional factor of -1, with an element of $S \times S$ viz.

$$\epsilon^{AB} = -(\epsilon^{-1})^{AB} = \begin{pmatrix} 0 & 1 \\ -1 & 0 \end{pmatrix}.$$

Although the minus sign may seem at this stage a trifle unnatural, its value will soon become apparent.

(2.2.3) Exercise. Show that

$$\epsilon^{AB} = o^A \iota^B - \iota^A o^B.$$

The convention on kernel letter of dual elements means that ϵ_{AB} can be regarded as an index-lowering-operator. For

$$[\xi, \eta] = \epsilon_{AB} \zeta^A \eta^B = (\epsilon_{AB} \zeta^A) \eta^B.$$

The bracketed quantity is clearly the dual of ξ^B, and so we have

$$\xi_B = \epsilon_{AB} \xi^A = \xi^A \epsilon_{AB}. \tag{2.2.2}$$

Since ϵ_{AB} is non-singular the map $\xi \mapsto [\xi, \ ]$ is an isomorphism, as claimed earlier. This equation has an important concomitant. Since ϵ_{AB} is non-singular we have

$$(\epsilon^{-1})^{BC} \xi_B = \xi^A \epsilon_{AB} (\epsilon^{-1})^{BC} = \xi^A \delta_A{}^C,$$

where $\delta_A{}^C$ is the spinor Kronecker delta or unit matrix. Note carefully the position of the indices. Evaluating the sum implicit in the right hand side we obtain

$$-\epsilon^{BC} \xi_B = \xi^A \delta_A{}^C = \xi^C,$$

or

$$\xi^C = \epsilon^{CB}\xi_B.\qquad(2.2.3)$$

Just as the metric tensor can be used to raise and lower indices in conventional tensor analysis so the symplectic structure can be used on spinors. Again note carefully the position of the indices in (2.2.2),(2.2.3). If the contraction is across adjacent 'northwest' and 'southeast' indices no minus signs occur. (This is why ϵ^{AB} was chosen to be **minus** the inverse of ϵ_{AB}.)

(2.2.4) Exercise. Use (2.2.2), (2.2.3) to show that

$$\epsilon_C{}^A = \delta_C{}^A = -\epsilon^A{}_C.$$

For this reason ϵ is usually used as the kernel letter for the Kronecker delta.

(2.2.5) Exercise. Show that for any ξ^{AB}, $\xi_A{}^A = -\xi^A{}_A$, and that $\epsilon^{AB}\epsilon_{AB} = \epsilon_A{}^A = 2$. Show further that $\eta^A\xi_A = 0$ if and only if η^A, ξ^A are proportional.

Next we consider a linear transformation applied to a spin basis

$$\tilde{o}^A = \alpha o^A + \beta \iota^A, \qquad \tilde{\iota}^A = \gamma o^A + \delta \iota^A.\qquad(2.2.4)$$

Since the same equations hold with indices lowered it is easy to see that $(\tilde{o}, \tilde{\iota})$ forms a spin basis if and only if $\alpha\delta - \beta\gamma = 1$, i.e., the transformation matrix $\begin{pmatrix} \alpha & \beta \\ \gamma & \delta \end{pmatrix}$ is in $SL(2,\mathcal{C})$. Since the condition that a linear transformation preserve spin bases is precisely the condition that it be a symplectomorphism we deduce immediately that $Sp(2)$ is isomorphic to $SL(2,\mathcal{C})$ as claimed earlier.

(2.2.6) Exercise. Demonstrate this directly by computing $\tilde{\epsilon}^{AB}$.

It should be obvious that the usual symmetrization ( ) and antisymmetrization [] operations can be applied to spinor suffices. Since the dimension of S is two, $\tau_{...[ABC]...} = 0$ for any multivalent spinor τ, for at least two of the bracketed indices must be equal. In particular we have the *Jacobi identity*

$$\epsilon_{A[B}\epsilon_{CD]} = 0 = \epsilon_{AB}\epsilon_{CD} + \epsilon_{AC}\epsilon_{DB} + \epsilon_{AD}\epsilon_{BC}.\qquad(2.2.5)$$

This is usually used in the form given by the next equation.

(2.2.7) LEMMA

Let $\tau_{...CD...}$ be a multivalent spinor. Then

$$\tau_{...AB...} = \tau_{...(AB)...} + \tfrac{1}{2}\epsilon_{AB}\tau_{...C}{}^{C}{}_{...}. \tag{2.2.6}$$

Proof: It is clearly sufficient to consider the case where τ has valence two. Then multiplying (2.2.5) with the CD indices raised into τ_{CD} gives

$$\epsilon_{AB}\tau_{C}{}^{C} - \tau_{AB} + \tau_{BA} = 0,$$

or

$$\tau_{[AB]} = \tfrac{1}{2}\epsilon_{AB}\tau_{C}{}^{C}.$$

Since $\tau_{AB} = \tau_{(AB)} + \tau_{[AB]}$ the result follows immediately.

This is a special case of a more general result, for which an elegant proof has been given by Penrose and Rindler, 1984.

(2.2.8) THEOREM

Any spinor $\tau_{A...F}$ is the sum of the totally symmetric spinor $\tau_{(A...F)}$ and (outer) products of ϵ's with totally symmetric spinors of lower valence.

Proof: The result is obviously true for spinors of valence 2, and we proceed by induction on n, the valence of τ. If the difference of two spinors α, β of the same valence is a sum of terms, each of which is an outer product of ϵ with a spinor of lower valence, we write $\alpha \bowtie \beta$. Clearly $\bowtie$ is an equivalence relation. Now

$$n\tau_{(ABC...F)} = \tau_{A(BC...F)} + \tau_{B(AC...F)} + \cdots + \tau_{F(BC...A)}.$$

Consider the difference between the first and any other term on the right, e.g.

$$\tau_{A(BC...F)} - \tau_{B(AC...F)}.$$

This is clearly skew-symmetric in AB and so by lemma 2.2.7 can be written as ϵ_{AB} times a spinor of valence $n-2$. Applying this to each term on the right after the first we find that all n terms on the right are equivalent to the first so that $\tau_{(ABC...F)} \bowtie \tau_{A(BC...F)}$. Thus if the result is true for valence $n-1$ then it is true for valence n, which proves the result.

Thus "only symmetric spinors matter", or more precisely every irreducible representation of $SL(2,C)$ can be realized by symplectomorphisms acting on symmetric spinors; see e.g., Naimark, 1964, for details.

Because S is a vector space over C one might naively assume that the complex conjugate of a spinor in S could be defined as an element of S. However there are (at least) two reasons why this is not so. Let us denote the operation of complex conjugation by a bar. If α, β are in S and c is in C the conjugate of $\alpha + c\beta$ would be $\bar{\alpha} + \bar{c}\bar{\beta}$ rather than $\bar{\alpha} + c\bar{\beta}$ and so complex conjugation is an *anti-isomorphism* and not an isomorphism. Secondly if complex conjugation were an isomorphism then 'real' and 'imaginary' valence one spinors could be distinguished. But it is clear from the discussion of transformations between spin bases that any valence one spinor can be mapped to any other by a symplectomorphism, so that a 'real' spinor would be isomorphic to an 'imaginary' one. It is clear therefore that complex conjugation must be regarded as an anti-isomorphism from S to a new vector space $\overline{S}$. To emphasize the fact that $\overline{S}$ is different from S, not only does the kernel letter of a spinor acquire a bar, but the suffix also gains a prime e.g.

$$\alpha^A \mapsto \overline{\alpha^A} = \bar{\alpha}^{A'},$$

with similar results for lowered indices. There is clearly some notational redundancy here, but apart from an exception given below it has become standard. As usual we regard complex conjugation applied twice as the identity map and identify $\overline{\overline{S}}$ with S, $\overline{\bar{\alpha}^{A'}} = \alpha^A$. The symplectic structure of S maps to a symplectic structure for $\overline{S}$. It is then apparent that the complex conjugate of S^* is the dual space of $\overline{S}$, which will be denoted $\overline{S}^*$. Note further that in component terms $\overline{\epsilon_{AB}}$ is always written as $\epsilon_{A'B'}$ and not as $\bar{\epsilon}_{A'B'}$. (This and the raised index version is the exception mentioned above.) We may now build a grand tensor algebra out of $S, S^*, \overline{S}, \overline{S}^*$. Because S and S^* are to be regarded as the 'same' vector space the order of unprimed indices matters, e.g.

$$\tau^{AB}{}_{CD} \neq \tau^A{}_C{}^B{}_D$$

in general. Similarly

$$\bar{\tau}^{A'B'}{}_{C'D'} \neq \bar{\tau}^{A'}{}_{C'}{}^{B'}{}_{D'}.$$

But since S and $\overline{S}$ are different vector spaces we do not need to distinguish between say $S \times \overline{S}$ and $\overline{S} \times S$. Thus all primed indices can be shuffled through unprimed ones and vice versa, e.g.

$$\tau_{AB'} = \tau_{B'A}.$$

Note however that there is a price to be paid for this convenience. We can no longer regard e.g., $\tau_{AB'}$ as a 2×2 matrix if we are allowed to set $\tau_{AB'} = \tau_{B'A}$. The only spinors that we interpret in this way are the Infeld-van der Waerden symbols defined in the next section. Therefore the possibility of index reshuffling must be rescinded for that particular example.

2.3 Spinors and vectors

A **Hermitian** spinor τ is one for which $\bar{\tau} = \tau$. For this to make sense τ must have as many primed as unprimed indices, and the relative positions of the primed indices must be the same as the relative positions of the unprimed ones. The simplest example is an element of $S \times \overline{S}$, e.g., $\tau^{AA'}$. Let $(o, \iota), (\bar{o}, \bar{\iota})$ be spin bases for S and $\overline{S}$. Then there exist scalars ξ, η, ζ, σ such that

$$\tau^{AA'} = \xi o^A \bar{o}^{A'} + \eta \iota^A \bar{\iota}^{A'} + \zeta o^A \bar{\iota}^{A'} + \sigma \iota^A \bar{o}^{A'}.$$

The assertion that $\tau^{AA'}$ is Hermitian is equivalent to the statement that ξ and η are real and that ζ and σ are complex conjugates.

Thus the set of Hermitian spinors $\tau^{AA'}$ forms a real vector space of dimension 4. In section 2.5 this will be identified with $T_p(M)$, the tangent space at a point in a 4-manifold M. Similarly the set of Hermitian spinors $\tau_{AA'}$ forms a real vector space of dimension 4, dual to the one above. In section 2.5 this will become the cotangent space $T_p^*(M)$. We denote the 4 real components of $\tau^{AA'}$ by τ^a. This is best understood as a relabelling $AA' \mapsto a$. (The explicit details are described below.) Every spin basis defines a tetrad of vectors $(l, n, m, \overline{m})$ via

$$
\begin{aligned}
l^a &= o^A \overline{o}^{A'}, & n^a &= \iota^A \overline{\iota}^{A'}, & m^a &= o^A \overline{\iota}^{A'}, & \overline{m}^a &= \iota^A \overline{o}^{A'}, \\
l_a &= o_A \overline{o}_{A'}, & n_a &= \iota_A \overline{\iota}_{A'}, & m_a &= o_A \overline{\iota}_{A'}, & \overline{m}_a &= \iota_A \overline{o}_{A'}.
\end{aligned}
\tag{2.3.1}
$$

Clearly l and n are real while m and $\overline{m}$ are complex conjugates. We may compute the various vector contractions as follows: evidently $l_a l^a$ is shorthand for $l_{AA'} l^{AA'}$ which expands to $o_A \overline{o}_{A'} o^A \overline{o}^{A'} = 0$.

(2.3.1) Exercise. Show that

$$
l_a n^a = l^a n_a = -m^a \overline{m}_a = -\overline{m}^a m_a = 1,
$$

while all other contractions vanish.

(2.3.2) DEFINITION

*A set of 4 vectors which possesses the properties of exercise 2.3.1 is called a **Newman-Penrose null tetrad**. This is usually abbreviated to **NP null tetrad** or even **null tetrad**.*

Another useful Hermitian spinor is

$$
g_{ABA'B'} = \epsilon_{AB} \epsilon_{A'B'}, \qquad g^{ABA'B'} = \epsilon^{AB} \epsilon^{A'B'},
$$

with tensor equivalents g_{ab}, g^{ab} which are obviously symmetric.

(2.3.3) Exercise. Use $\epsilon_{AB} = o_A \iota_B - \iota_A o_B$ to show that

$$
g_{ab} = 2l_{(a} n_{b)} - 2m_{(a} \overline{m}_{b)}, \qquad g^{ab} = 2l^{(a} n^{b)} - 2m^{(a} \overline{m}^{b)}.
\tag{2.3.2}
$$

Hence, or otherwise, show that

$$l^a = g^{ab}l_b, \qquad l_a = g_{ab}l^b, \qquad g^{ab}g_{bc} = \delta_c{}^a \quad etc.$$

Thus g has all the properties of a metric tensor. To make this even more explicit define a tetrad of vectors

$$e_{\hat{0}} = 2^{-1/2}(l+n), \qquad e_{\hat{1}} = 2^{-1/2}(m+\overline{m}),$$
$$e_{\hat{2}} = 2^{-1/2}i(m-\overline{m}), \qquad e_{\hat{3}} = 2^{-1/2}(l-n). \qquad (2.3.3)$$

(2.3.4) Exercise. Show that $(e_{\hat{a}})$ form an **orthonormal tetrad**, i.e.,

$$g_{\hat{a}\hat{b}} = \text{diag}(1,-1,-1,-1).$$

Thus the existence of a spinor structure fixes the signature of spacetime to be $(+ - - -)$.

We can spell out the correspondence $\tau^{AA'} \mapsto \tau^a$ explicitly by writing

$$\epsilon_{AB}\epsilon_{A'B'} = g_{\hat{a}\hat{b}}\sigma^{\hat{a}}{}_{AA'}\sigma^{\hat{b}}{}_{BB'}, \qquad (2.3.4)$$

where the σ's are the **Infeld-van der Waerden symbols**. (Note that the primed index is always last. See the remark at the end of the last section.) A suitable choice is

$$\sigma_{\hat{a}}{}^{AA'} = 2^{-1/2}\sigma_{\hat{a}} \qquad a = 0,1,2,3, \qquad (2.3.5)$$

where the $\sigma_{\hat{a}}$ are the Pauli matrices of section 2.1. (It is a tedious exercise to verify that they satisfy (2.3.4).)

(2.3.5) Exercise. Raise and lower indices to show that

$$\sigma^{\hat{a}}{}_{AA'} = \lambda_{\hat{a}}2^{-1/2}\sigma_{\hat{a}}, \qquad \text{where} \quad \lambda_{\hat{2}} = -1, \quad \lambda_{\hat{a}} = 1 \text{ otherwise.} \qquad (2.3.6)$$

(The summation convention does not apply to equation (2.3.6).) If u is any spacetime vector we may write

$$u = u^a e_{\hat{a}}{}^a = u^{AA'}\sigma^{\hat{a}}{}_{AA'}e_{\hat{a}}{}^a,$$

so that

$$u^{AA'} = \sigma_{\hat{a}}{}^{AA'}u^{\hat{a}}.$$

This is the identification discussed in section 2.1. Apart from Appendix A the Infeld-van der Waerden symbols will not appear again!

We now proceed to the spacetime interpretation of spinors. Clearly as we have already seen, every univalent spinor κ^A defines a real null vector k^a via $k^a = \kappa^A \bar{\kappa}^{A'}$. This is a special case of a more general result.

(2.3.6) THEOREM

Every non-vanishing real null vector k^a can be written in one or other of the forms

$$k^a = \pm \kappa^A \bar{\kappa}^{A'}. \tag{2.3.7}$$

Proof: It is clear that k^a defined via (2.3.7) is real and null. Suppose conversely that $k^a = \omega^{AA'}$ is real and null. The nullity condition is

$$\epsilon_{AB} \epsilon_{A'B'} \omega^{AA'} \omega^{BB'} = 0,$$

which says that the 2×2 matrix $\omega^{AA'}$ has vanishing determinant, so that the rows/columns are linearly dependent. This means that there exist univalent spinors κ, λ such that

$$\omega^{AA'} = \kappa^A \bar{\lambda}^{A'}.$$

The reality condition is

$$\kappa^A \bar{\lambda}^{A'} = \lambda^A \bar{\kappa}^{A'}.$$

Multiplying by κ_A implies, assuming $\kappa \neq 0$, that $\kappa_A \lambda^A = 0$. In other words λ is proportional to κ. Renormalizing κ now gives (2.3.4) where the sign is that of the (real) proportionality constant.

Every univalent spinor κ^A defines a null vector k^a. However $e^{i\theta} \kappa^A$ defines the same null vector for real θ and so there is some additional 'phase' information in κ. To extract this information we complete κ to a spin basis (κ, μ) where μ is unique up to an additive multiple of κ. Now

$$s^a = 2^{-1/2} (\kappa^A \bar{\mu}^{A'} + \mu^A \bar{\kappa}^{A'}),$$

is a unit spacelike vector orthogonal to k^a, and is unique up to an additive multiple of k^a. However the 2-plane in the tangent space spanned by k^a, s^a is unique. Another real unit spacelike vector orthogonal to k^a is

$$t^a = 2^{-1/2} i (\kappa^A \bar{\mu}^{A'} - \mu^A \bar{\kappa}^{A'}).$$

Together s, t span a spacelike 2-surface (in the tangent space) orthogonal to k. Now suppose we make a phase change to κ, i.e., $\kappa \mapsto e^{i\theta}\kappa$. Clearly $\bar{\kappa} \mapsto e^{-i\theta}\bar{\kappa}$ and since (κ, μ) form a spin basis, i.e., $\kappa_A \mu^A = 1$, it follows that $\mu \mapsto e^{-i\theta}\mu$. It follows directly from the definitions that

$$s^a \mapsto \cos 2\theta s^a + \sin 2\theta t^a, \qquad t^a \mapsto -\sin 2\theta s^a + \cos 2\theta t^a.$$

Penrose has suggested the geometrical interpretation of a univalent spinor as a flag. The 'flagpole' direction of κ^A is defined to be parallel to the direction of k^a, while the 'flag' itself lies in the two-plane spanned by k^a and s^a. Thus the flagpole lies in the plane of the flag. Changing the phase of κ^A by θ leaves the flagpole invariant but rotates the flag plane around the pole by an angle 2θ as measured in the spacelike 2-plane orthogonal to the pole spanned by s^a and t^a. Note that a phase change by π leaves the flag plane invariant, but reverses the sign of κ^A. This is another manifestation of the fact that the map $SL(2, \mathcal{C}) \to L(4)$ is $(2, 1)$.

So far our considerations have been local; we have considered spinors and tensors defined at a point. Suppose now that it is possible to define spinor fields in some region of spacetime. We must then ask whether this places any restriction on the spacetime? Every non-zero null vector must be either future pointing or past pointing. We see immediately from theorem 2.3.6 that we have a consistent method for distinguishing the two possibilities; a null vector will be said to be *future pointing* if the $+$ sign occurs in the decomposition (2.3.7) and *past pointing* otherwise. Thus our spacetime must be time-orientable. Further the effect of a phase change on a flag can be used to define a space-orientation in the spacetime: the flag plane rotation corresponding to $\theta > 0$ is defined to be right-handed. Then the orthonormal tetrad (2.3.3) has the standard orientation, i.e., $e_{\hat{0}}$ is future point-ing and $(e_{\hat{1}}, e_{\hat{2}}, e_{\hat{3}})$ form a right-handed triad of vectors. (Such a tetrad will be called a *Minkowski tetrad.*) Thus a necessary condition for the existence of spinor fields is that spacetime be ori-entable. In a non-compact spacetime the existence of 4 continuous vector fields constituting a Minkowski tetrad at each point is not only necessary but also sufficient for the existence of spinor fields. (A proof is given in Geroch, 1968.) Most physicists would accept these conditions as being very reasonable and we shall henceforth assume them without further comment.

2.4   The Petrov classification

The Petrov classification of zero rest mass fields rests on two pieces of algebra. It will be recalled that in tensor algebra the *alternating tensor* is defined by

$$\epsilon_{abcd} = \epsilon_{[abcd]}, \qquad \epsilon_{abcd}\epsilon^{abcd} = -24, \qquad \epsilon_{\hat{0}\hat{1}\hat{2}\hat{3}} = 1, \qquad (2.4.1)$$

where the last equation refers to the components with respect to a Minkowski orthonormal tetrad such as (2.3.3). Clearly its spinor equivalent will have four primed and four unprimed indices, will change sign under the appropriate index interchanges and will be constructed out of the ϵ_{AB}'s. Some experimentation reveals that essentially the only candidate is

$$\epsilon_{abcd} = \epsilon_{ABCDA'B'C'D'} = \lambda(\epsilon_{AB}\epsilon_{CD}\epsilon_{A'C'}\epsilon_{B'D'} - \epsilon_{AC}\epsilon_{BD}\epsilon_{A'B'}\epsilon_{C'D'}),$$

where λ is a constant. The second equation (2.4.1) now fixes $\lambda^2 = -1$ so that only the sign of λ needs to be determined. This is a straightforward calculation using the last equation (2.4.1) and (2.3.3) and gives $\lambda = i$.

(2.4.1) Exercise. Carry out the calculations outlined above.

As we saw in section 2.2 it is possible to decompose any spinor into its totally symmetric part plus sums of products of ϵ_{AB}'s with lower valence symmetric spinors. The following theorem shows how totally symmetric spinors can themselves be factorized.

(2.4.2) THEOREM

Suppose $\tau_{AB...C}$ is totally symmetric. Then there exist univalent spinors $\alpha_A, \beta_B, \ldots, \gamma_C$ such that

$$\tau_{AB...C} = \alpha_{(A}\beta_B \cdots \gamma_{C)}.$$

*The $\alpha, \beta, \ldots, \gamma$ are called the **principal spinors** of τ. The corresponding real null vectors are called the **principal null directions** of τ.*

Proof: Let τ have valence n and let $\xi^A = (x, y)$. Define

$$\tau(\xi) = \tau_{AB...C}\xi^A\xi^B \cdots \xi^C.$$

This is clearly a homogeneous polynomial of degree n in the (complex) x, y and so can be factorized completely

$$\tau(\xi) = (\alpha_0 x - \alpha_1 y)(\beta_0 x - \beta_1 y) \ldots (\gamma_0 x - \gamma_1 y),$$

which proves the result since x and y are arbitrary. Notice that neither the order nor possible factors in the $\alpha, \beta, \ldots, \gamma$ are specifiable.

As a first example we consider the totally skew Maxwell tensor F_{ab} for the electromagnetic field. This has a spinor equivalent $F_{ABA'B'} = -F_{BAB'A'}$. We define

$$\phi_{AB} = \tfrac{1}{2}F_{ABC'}{}^{C'} = \phi_{BA},$$

where the last equality follows from the symmetry properties of F. Using lemma 2.2.7 we have immediately

$$F_{ABA'B'} = F_{AB(A'B')} + \phi_{AB}\epsilon_{A'B'},$$

and a second application gives

$$F_{ABA'B'} = F_{(AB)(A'B')} + \phi_{AB}\epsilon_{A'B'} + \epsilon_{AB}\bar{\phi}_{A'B'}.$$

The first term on the right vanishes because of the symmetry properties of F so that

$$F_{ABA'B'} = \phi_{AB}\epsilon_{A'B'} + \epsilon_{AB}\bar{\phi}_{A'B'}.$$

If we define the **dual of** F as $F^*_{ab} = \frac{1}{2}\epsilon_{ab}{}^{cd}F_{cd}$ then it is straightforward to show that

$$F_{ab} + iF^*_{ab} = 2\phi_{AB}\epsilon_{A'B'}, \tag{2.4.2}$$

so that ϕ_{AB} and F_{ab} are fully equivalent.

(2.4.3) Exercise. Derive equation (2.4.2).

We now use theorem 2.4.2 to decompose ϕ_{AB} as

$$\phi_{AB} = \alpha_{(A}\beta_{B)}. \tag{2.4.3}$$

There are two possibilities here. If α and β are proportional then α is called a **repeated principal spinor** of ϕ, and ϕ is said to be **algebraically special** or **null**, or of **type N**. The corresponding real null vector $\alpha_a = \alpha_A\bar{\alpha}_{A'}$ is called a **repeated principal null direction (rpnd)**. If α and β are not proportional ϕ is said to be **algebraically general** or of **type I**.

(2.4.4) Exercise. Show that ϕ is null if and only if

$$\tfrac{1}{2}F_{ab}F^{ab} = |\mathbf{B}|^2 - |\mathbf{E}|^2 = 0, \qquad -\tfrac{1}{2}\epsilon_{abcd}F^{ab}F^{cd} = \mathbf{E}.\mathbf{B} = 0.$$

Suggest an explicit example of a null electromagnetic field.

As a second example we consider gravity. We use a result, to be derived later, that the Weyl tensor $C_{abcd} = C_{[ab][cd]} = C_{cdab}$, to be defined later, can be written as

$$C_{abcd} = C_{ABCDA'B'C'D'}$$
$$= \Psi_{ABCD}\epsilon_{A'B'}\epsilon_{C'D'} + \overline{\Psi}_{A'B'C'D'}\epsilon_{AB}\epsilon_{CD}, \tag{2.4.4}$$

where Ψ_{ABCD} is totally symmetric. The **dual of the Weyl tensor** is

$$C^*_{abcd} = \tfrac{1}{2}\epsilon_{cd}{}^{ef}C_{abef},$$

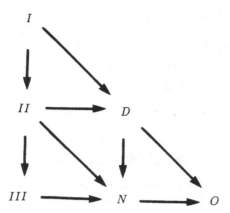

Fig. 2.4.1   A Penrose diagram of the Petrov classification.   The arrows indicate increasing specialization of the solutions.

and

$$C_{abcd} + iC^*{}_{abcd} = 2\Psi_{ABCD}\epsilon_{A'B'}\epsilon_{C'D'},$$

so that C_{abcd} and Ψ_{ABCD} are fully equivalent. Theorem 2.4.2 now implies that

$$\Psi_{ABCD} = \alpha_{(A}\beta_B\gamma_C\delta_{D)}. \qquad (2.4.5)$$

There are six cases to be distinguished here.

Type I or $\{1,1,1,1\}$. None of the four principal null directions coincide. This is the *algebraically general* case.

Type II or $\{2,1,1\}$. Two directions coincide. This and all subsequent cases are *algebraically special*.

Type D or $\{2,2\}$. There are two (different) pairs of repeated principal null directions.

Type III or $\{3,1\}$. Three principal null directions coincide.

Type N or $\{4\}$. All four principal null directions coincide.

This is known as the ***Petrov classification***, and the increasing specialization is succintly represented by the arrows in a ***Penrose diagram***, figure 2.4.1. (If spacetime is empty, O is flat space!) The description of this classification by any other approach is extremely messsy. Ψ has two scalar invariants

$$I = \Psi_{ABCD}\Psi^{ABCD}, \qquad J = \Psi_{AB}{}^{CD}\Psi_{CD}{}^{EF}\Psi_{EF}{}^{AB}.$$

The condition for type II is that $I^3 = 6J^2$ and for type III that $I = J = 0$. Type D is fixed by

$$\Psi_{PQR(A}\Psi_{BC}{}^{PQ}\Psi^{R}{}_{DEF)} = 0,$$

while type N is fixed by

$$\Psi_{(AB}{}^{EF}\Psi_{CD)EF} = 0.$$

The tensor equivalents are even more messy. See e.g., Pirani, 1965, for the details.

(2.4.5) Problem. A skew symmetric (possibly complex) valence 2 tensor or *bivector* F_{ab} is said to be *self-dual* if $F^{*}{}_{ab} = iF_{ab}$ and *anti-self-dual* if $F^{*}{}_{ab} = -iF_{ab}$. Investigate the spinorial equivalents and show that: a) every bivector is the sum of a self-dual and an anti-self-dual bivector, b) the product of a self-dual and an anti-self-dual bivector vanishes. Can this be generalized to 4-index tensors with Weyl symmetries?

(2.4.6) Problem. Show that the space of valence 1 spinors at a point can be mapped onto a (double) null cone and hence onto a unit 2-sphere if $+k^a$ and $-k^a$ are regarded as the same direction. Show that the Weyl spinor Ψ_{ABCD} defines a finite set of points on the sphere, and examine the various Petrov types. A Friedmann-Robertson-Walker spacetime is isotropic at every point. What can be said about the Weyl tensor (and gravitational radiation) in such spacetimes? Determine the Petrov type of a Schwarzschild spacetime from the properties:

a) vacuum apart from a static point source,
b) axially symmetric about the spatial direction of the source,
c) time-symmetric.

2.5 Spinor analysis

So far attention has been focussed on the algebra of spinors, i.e., the spinor structure at a point. Given a spinor structure at each point of spacetime one wants them to mesh together smoothly. Consider first the possibility of a Lie derivative of a spinor field. If such a derivative existed we could compute e.g., $\mathcal{L}_X \epsilon_{AB}$. This object would have 2 spinor indices and be skew, and hence would be proportional to ϵ_{AB}. It follows that $\mathcal{L}_X g_{ab}$ would be proportional to g_{ab}, i.e., X would have to be a conformal Killing vector field. Thus Lie differentiation of spinors is not possible in general. One way to define a spinor covariant derivative would be to note that spinors can be interpreted in terms of tensors, and to extend consistently the tensor covariant derivative. The viewpoint adopted here though is that spinors may well be more fundamental than tensors and so an axiomatic approach will be adopted. In the definition given below not all of the axioms are entirely independent, but this slightly redundant description is clearer than a minimal definition.

(2.5.1) DEFINITION

*Let θ, ϕ and ψ be spinor fields such that θ and ϕ have the same valence. The **spinor covariant derivative** is defined axiomatically as a map $\nabla_x = \nabla_{XX'}: \theta_{...} \mapsto \theta_{...;XX'}$ such that*

i) $\nabla_x(\theta + \phi) = \nabla_x \theta + \nabla_x \phi$,

ii) $\nabla_x(\theta\psi) = (\nabla_x \theta)\psi + \theta(\nabla_x \psi)$,

iii) $\psi = \nabla_x \theta$ implies $\bar{\psi} = \nabla_x \bar{\theta}$,

iv) $\nabla_x \epsilon_{AB} = \nabla_x \epsilon^{AB} = 0$,

v) ∇_x commutes with any index substitution not involving X, X',

vi) $\nabla_x \nabla_y f = \nabla_y \nabla_x f$ for all scalars f,

vii) for any derivation D acting on spinor fields there exists a spinor $\xi^{XX'}$ such that $D\psi = \xi^{XX'} \nabla_{XX'} \psi$ for all ψ.

(Here a *derivation* is a linear map acting on spinor/tensor fields which obeys the Leibniz rule ii) and annihilates constant scalar fields.)

All of these conditions are conventional except for vii). This is needed to ensure that $\xi^{XX'}$ is a (possibly complex) tangent vector in the spacetime. This identifies the 4-dimensional vector space of section 2.3 with the tangent space $T_p(M)$ at a point $p \in M$, and its dual with the cotangent space $T_p^*(M)$. (Up to now we have constructed merely a 4-dimensional vector space at each point.) Note that condition vi) is equivalent to the absence of torsion.

(2.5.2) THEOREM

∇_x *exists and is unique.*

Proof: The somewhat complicated proof is given in section 4.4 of Penrose and Rindler, 1984.

∇_x has two desirable properties. Firstly if $\nabla_x \kappa^A = 0$ then the corresponding flagpole and flagplane are parallelly transported. Secondly condition iv) implies that $\nabla_x g_{ab} = 0$ so that, acting on tensors, ∇_x is the metric covariant derivative.

In tensor analysis the most efficient way to compute the curvature is usually via a tetrad formalism. One introduces a tetrad of vectors $e_{\hat{a}}{}^a$ and a dual basis of covectors $e^{\hat{a}}{}_a$, i.e., $e_{\hat{a}}{}^a e^{\hat{a}}{}_b = \delta_b{}^a$. Here hatted indices label the vectors, while unhatted vectors label the components with respect to some arbitrarily chosen basis. The notation is the same as that for Minkowski tetrads introduced earlier.

The Christoffel symbols are replaced by the **Ricci rotation coefficients** defined by

$$\Gamma_{\hat{a}\hat{b}\hat{c}} = e_{\hat{a}a} \nabla_{\hat{c}} e_{\hat{b}}{}^a \equiv e_{\hat{a}a} e_{\hat{c}}{}^c \nabla_c e_{\hat{b}}{}^a = -e_{\hat{b}}{}^a \nabla_{\hat{c}} e_{\hat{a}a}. \qquad (2.5.1)$$

For a torsion-free metric connection the **Riemann curvature tensor** of chapter 1 can be computed from the **Ricci identity**

$$u^a R_{abcd} = 2\nabla_{[c} \nabla_{d]} u_b,$$

for any vector u^a. Now u_b can be written as a linear combination of the e's viz. $u_b = u_{\hat{b}} e^{\hat{b}}{}_b$. Here the $u_{\hat{b}}$'s are *scalars*. (Note the absence of any tensorial index in $u_{\hat{b}}$.) Further it is easy to show that $\nabla_{[c} \nabla_{d]}$ is a derivation with the additional property that it annihilates all scalar fields, not just the constant ones. Thus

$$u^a R_{abcd} = 2\nabla_{[c}\nabla_{d]}(u_{\hat{b}}e^{\hat{b}}{}_b)$$
$$= 2u_{\hat{b}}\nabla_{[c}\nabla_{d]}e^{\hat{b}}{}_b$$
$$= 2(u^a e_{\hat{b}a})\nabla_{[c}\nabla_{d]}e^{\hat{b}}{}_b.$$

It follows that

$$R_{abcd} = 2e_{\hat{b}a}\nabla_{[c}\nabla_{d]}e^{\hat{b}}{}_b. \tag{2.5.2}$$

Although (2.5.2) may look unfamiliar it leads directly to its spinor equivalent.

We introduce a spinor dyad $\epsilon_{\hat{A}}{}^A$ and its symplectic dual $\epsilon^{\hat{A}}{}_A$. The defining formula is

$$\epsilon_{\hat{A}}{}^A \epsilon^{\hat{A}}{}_B = \epsilon^A{}_B, \tag{2.5.3}$$

and a suitable choice is $\epsilon_{\hat{0}}{}^A = o^A$, $\epsilon_{\hat{1}}{}^A = \iota^A$, $\epsilon^{\hat{0}}{}_A = -\iota_A$, $\epsilon^{\hat{1}}{}_A = o_A$, consistent with (2.2.2), (2.2.3) and exercise 2.2.4. We could now introduce the **spinor Ricci rotation coefficients**

$$\Gamma_{\hat{A}\hat{B}\hat{C}\hat{C}'} = \epsilon_{\hat{A}A}\epsilon_{\hat{C}}{}^C \epsilon_{\hat{C}'}{}^{C'}\nabla_{CC'}\epsilon_{\hat{B}}{}^A, \tag{2.5.4}$$

although a more useful notation will be introduced in the next section. The spinor equivalent of (2.5.2) is

$$R_{ABCDA'B'C'D'} = 2\epsilon_{\hat{B}A}\epsilon_{\hat{B}'A'}\nabla_{[c}\nabla_{d]}(\epsilon^{\hat{B}}{}_B \epsilon^{\hat{B}'}{}_{B'})$$
$$= 2\epsilon_{\hat{B}A}\epsilon_{\hat{B}'A'}\epsilon^{\hat{B}'}{}_{B'}\nabla_{[c}\nabla_{d]}\epsilon^{\hat{B}}{}_B + c.c.$$
$$= 2\epsilon_{\hat{B}A}\epsilon_{A'B'}\nabla_{[c}\nabla_{d]}\epsilon^{\hat{B}}{}_B + c.c. \tag{2.5.5}$$

Here c.c. denotes the complex conjugate and the fact that $\nabla_{[c}\nabla_{d]}$ is a derivation and equation (2.5.3) have been used.

Next we start to evaluate the right hand side of (2.5.5). We first define

$$\Box_{CD} = \epsilon^{C'D'}\nabla_{[CC'}\nabla_{DD']}$$
$$= \tfrac{1}{2}\epsilon^{C'D'}(\nabla_{CC'}\nabla_{DD'} - \nabla_{DD'}\nabla_{CC'})$$
$$= \tfrac{1}{2}(\nabla_{CC'}\nabla_D{}^{C'} + \nabla_{DD'}\nabla_C{}^{D'})$$
$$= \nabla_{C'(C}\nabla_{D)}{}^{C'}.$$

Similarly we define

$$\square_{C'D'} = \nabla_{C(C'}\nabla_{D')}{}^{C}.$$

(2.5.3) Exercise. Let D be one of the operators $\nabla_{[a}\nabla_{b]}$, $\square_{AB}$, $\square_{A'B'}$. Suppose that the tensors (spinors) U_1, U_2 have the same valence, while U, V have arbitrary valence. Show that

$$D(U_1 + U_2) = DU_1 + DU_2, \qquad D(UV) = (DU)V + UDV.$$

Show further that $D\phi = 0$ for any scalar ϕ.

Now by successively applying lemma 2.2.7 to the index pairs CD and $C'D'$ in the following equation we obtain

$$\begin{aligned}
2\nabla_{[c}\nabla_{d]} &= \nabla_{CC'}\nabla_{DD'} - \nabla_{DD'}\nabla_{CC'} \\
&= \nabla_{(C'(C}\nabla_{D)D')} + \tfrac{1}{2}\epsilon_{C'D'}\nabla_{E'(C}\nabla_{D)}{}^{E'} + \\
&\quad \tfrac{1}{2}\epsilon_{CD}\nabla_{E(C'}\nabla_{D')}{}^{E} + \tfrac{1}{4}\epsilon_{CD}\epsilon_{C'D'}\nabla_{EE'}\nabla^{EE'} - \\
&\quad \nabla_{(D'(D}\nabla_{C)C')} - \tfrac{1}{2}\epsilon_{D'C'}\nabla_{E'(D}\nabla_{C)}{}^{E'} - \\
&\quad \tfrac{1}{2}\epsilon_{DC}\nabla_{E(D'}\nabla_{C')}{}^{E} - \tfrac{1}{4}\epsilon_{DC}\epsilon_{D'C'}\nabla_{EE'}\nabla^{EE'}.
\end{aligned}$$

Thus

$$2\nabla_{[c}\nabla_{d]} = \epsilon_{C'D'}\square_{CD} + \epsilon_{CD}\square_{C'D'}. \tag{2.5.6}$$

Now consider a typical term on the right hand side of (2.5.5) e.g.

$$\epsilon_{\dot{B}A}\square_{CD}\epsilon^{\dot{B}}{}_{B} \ .$$

This is obviously symmetric in CD and if one applies $\square_{CD}$ to $\epsilon_{\dot{B}A}\epsilon^{\dot{B}}{}_{B} = 0$, it can be seen to be symmetric in AB. We now use lemma 2.2.7 to decompose it into irreducible symmetric spinors multiplied by ϵ's. Because of the symmetry properties there will be no terms consisting of a symmetric valence 2 spinor multiplied by an ϵ, and we need only apply lemma 2.2.7 to the index pairs DB, CA. Thus we may write

$$\epsilon_{\dot{B}A}\square_{CD}\epsilon^{\dot{B}}{}_{B} = \Psi_{ABCD} - 2\Lambda\epsilon_{(A(C}\epsilon_{D)B)}. \tag{2.5.7}$$

Here

$$\Psi_{ABCD} = \epsilon_{\dot{B}A}\square_{(CD}\epsilon^{\dot{B}}{}_{B)} = \Psi_{(ABCD)}, \tag{2.5.8}$$

and multiplying (2.5.7) by $\epsilon^{BC}\epsilon^{DA}$ gives

$$\Lambda = (1/6)\epsilon_{\hat{B}A}\Box^{AB}\epsilon^{\hat{B}}{}_{B}. \tag{2.5.9}$$

Also

$$\epsilon_{\hat{B}A}\Box_{C'D'}\epsilon^{\hat{B}}{}_{B} = \Phi_{ABC'D'}, \tag{2.5.10}$$

is symmetric on AB and on $C'D'$. Combining all of this into (2.5.5) gives

$$R_{ABCDA'B'C'D'} = \epsilon_{A'B'}\epsilon_{C'D'}[\Psi_{ABCD} - 2\Lambda\epsilon_{(A(C}\epsilon_{D)B)}]+ \atop \epsilon_{A'B'}\epsilon_{CD}\Phi_{ABC'D'} + c.c. \tag{2.5.11}$$

(2.5.4) Exercise. Show that the identity $R_{a[bcd]} = 0$ or equivalently $\epsilon^{abcd}R_{bcde} = 0$ implies that Λ is real and $\Phi_{ABA'B'}$ is Hermitian.

Thus $\Phi_{ABA'B'}$ and Λ have tensor analogues Φ_{ab}, Λ. Contracting (2.5.11) gives

$$R_{ABA'B'} = -2\Phi_{ABA'B'} + 6\Lambda\epsilon_{AB}\epsilon_{A'B'},$$

or

$$R_{ab} = -2\Phi_{ab} + 6\Lambda g_{ab}, \tag{2.5.12}$$

and a further contraction gives

$$R = 24\Lambda,$$

since Φ is clearly trace-free. Thus

$$\Lambda = R/24, \qquad \Phi_{ab} = -\tfrac{1}{2}(R_{ab} - \tfrac{1}{4}Rg_{ab}). \tag{2.5.13}$$

The remaining terms in (2.5.11) form a Hermitian spinor which we write as

$$C_{abcd} = \Psi_{ABCD}\epsilon_{A'B'}\epsilon_{C'D'} + \overline{\Psi}_{A'B'C'D'}\epsilon_{AB}\epsilon_{CD}, \tag{2.5.14}$$

which is known as the **Weyl curvature tensor**. By shifting around the terms in (2.5.11) one eventually obtains

$$C_{abcd} = R_{abcd} + 4\Phi_{[a[c}g_{d]b]} - 4\Lambda g_{[a[c}g_{d]b]} \atop = R_{abcd} - 2R_{[a[c}g_{d]b]} + \tfrac{1}{3}Rg_{[a[c}g_{d]b]}. \tag{2.5.15}$$

C_{abcd} has the same symmetries as R_{abcd} and is in addition trace-free. It represents the part of R_{abcd} not determined by local matter. It is often referred to as gravitational radiation although it may be time-independent.

(2.5.5) Exercise. Show that for an arbitrary spinor ξ_A

$$\Box_{AB}\xi_C = \Psi_{ABCD}\xi^D - 2\Lambda\xi_{(A}\epsilon_{B)C}, \qquad \Box_{(AB}\xi_{C)} = \Psi_{ABCD}\xi^D,$$

$$\Box_{AB}\xi^B = -3\Lambda\xi_A, \qquad \Box_{A'B'}\xi_C = \xi^D\Phi_{CDA'B'}.$$

(These are the spinor forms for the Ricci identities.)

(2.5.6) Exercise. The Bianchi Identities are $\nabla^d R^*{}_{abcd} = 0$ where $R^*{}_{abcd} = \frac{1}{2}\epsilon_{cd}{}^{ef}R_{abef}$. Obtain the spinor equivalents

$$-iR^*{}_{abcd} = \epsilon_{AB}\epsilon_{CD}\overline{\Psi}_{A'B'C'D'} - \epsilon_{A'B'}\epsilon_{C'D'}\Psi_{ABCD}+$$
$$\epsilon_{A'B'}\epsilon_{CD}\Phi_{ABC'D'} - \epsilon_{AB}\epsilon_{C'D'}\Phi_{CDA'B'}+$$
$$2\Lambda(\epsilon_{A'B'}\epsilon_{C'D'}\epsilon_{D(A}\epsilon_{B)C} - \epsilon_{AB}\epsilon_{CD}\epsilon_{D'(A'}\epsilon_{B')C'}),$$

$$\nabla^D{}_{C'}\Psi_{ABCD} = \nabla^{D'}{}_{(C}\Phi_{AB)C'D'},$$

$$\nabla^{BB'}\Phi_{ABA'B'} = -3\nabla_{AA'}\Lambda.$$

One may well ask why such clumsy formulae could be useful. Consider first a spin 1 zero rest mass field in vacuum, e.g., electromagnetism. Maxwell's equations are equivalent to

$$\nabla^a(F_{ab} + iF^*{}_{ab}) = 0,$$

or from (2.4.2)

$$\nabla^{AA'}\phi_{AB} = 0. \tag{2.5.16}$$

Similarly consider a spin 2 zero rest mass field, i.e., gravity. The Bianchi identities are

$$\nabla_{[e}R_{cd]ab} = 0,$$

or

$$\nabla^d R^*{}_{abcd} = 0.$$

It can be shown that in vacuum this implies

$$\nabla^{DD'}\Psi_{ABCD} = 0. \qquad (2.5.17)$$

(2.5.7) Problem. Show from exercise 2.5.6 that in a vacuum spacetime

$$\nabla_F{}^{A'}\Psi^F{}_{BCD} = 0.$$

By differentiating a second time, (use $\nabla_{AA'}$,) show that

$$\Box\Psi_{ABCD} = 6\Psi_{EF(AB}\Psi_{CD)}{}^{EF},$$

where $\Box = \nabla_{AA'}\nabla^{AA'}$ is the usual wave operator. What Petrov classes admit exact plane gravitational waves

$$\Box\Psi_{ABCD} = 0?$$

What can be said about linearized gravitational waves on a given background?

In fact zero rest mass fields of spin $s = \frac{1}{2}n$ where n is an integer or zero have irreducible representations in terms of totally symmetric spinors $\phi_{AB...C}$ of valence n satisfying

$$\nabla^{CC'}\phi_{AB...C} = 0. \qquad (2.5.18)$$

If $n \geq 3$ there is a consistency condition, the *Buchdahl constraint* that must be satisfied.

(2.5.8) Problem. Let $\phi_{ABCD...}$ be a valence $2s$ totally symmetric spinor, representing a zero rest mass spin $2s$ field. Show that for $s \geq 3/2$ the *Buchdahl constraint*, Buchdahl, 1958

$$\Psi^{ABC}{}_{(D}\phi_{EF...)ABC} = 0,$$

must hold. Investigate what happens for $s = 1$, $s = 2$.

(2.5.9) Problem. Let o^A be a constant spinor in Minkowski spacetime. Let χ be a (complex) scalar field satisfying $\Box\chi = 0$. Show that

$$\phi_{AB...C} = \nabla_A{}^{A'}\nabla_B{}^{B'} \ldots \nabla_C{}^{C'}(\chi\bar{o}_{A'}\bar{o}_{B'}\ldots\bar{o}_{C'}),$$

is totally symmetric and

$$\nabla^{AA'}\phi_{AB...C} = 0.$$

Thus every χ satisfying $\Box\chi = 0$ generates a spin s zero rest mass field. In fact the converse is true in any simply connected region. See Penrose, 1965, especially p.165. (χ is a Hertz-Bromwich-Whittaker-Debye-Penrose potential.)

Besides the examples given above there are some theories which are actually simpler in a spinor formulation. Let ϕ_A be an arbitrary spinor field. The following identity is obvious

$$2\phi_A\nabla_{EE'}\phi_B \equiv 2\phi_{(A}\nabla_{|EE'|}\phi_{B)} + 2\phi_{[A}\nabla_{|EE'|}\phi_{B]}$$
$$= \nabla_{EE'}(\phi_A\phi_B) + \epsilon_{AB}\phi_C\nabla_{EE'}\phi^C, \quad (2.5.19)$$

where the indices between the vertical lines are to be excluded from the symmetry operations, and lemma 2.2.7 has been used in the last line. Next define the vectors

$$M_e = \phi_A\nabla_{EE'}\phi^A, \qquad R_{AA'} = \phi_A\nabla_{BA'}\phi^B, \quad (2.5.20)$$

and the (anti-self-dual null) bivector

$$F_{ab} = \phi_A\phi_B\epsilon_{A'B'}. \quad (2.5.21)$$

It is straightforward to establish the identities

$$F_{ab}\nabla_e F_c{}^b = F_{ac}M_e, \quad (2.5.22)$$

$$\nabla_b F_a{}^b + M_a = 2R_a. \quad (2.5.23)$$

(2.5.10) Exercise. A ***Dirac-Weyl spinor*** satisfies

$$\nabla_{AA'}\phi^A = 0. \quad (2.5.24)$$

Show that $R_a = 0$, and hence that the tensor equivalent of (2.5.24) is

$$F_{ab}\nabla_e F_c{}^e + F_{ae}\nabla_c F_b{}^e = 0,$$

a considerably more complicated equation.

(2.5.11) Problem. Dirac's equation couples two spinor fields via

$$\nabla_{A'}{}^{A}\phi_A = \mu\chi_{A'}, \qquad \nabla_A{}^{A'}\chi_{A'} = \mu\phi_A, \qquad (2.5.25)$$

where μ is a real constant proportional to the mass of the field. Show that

$$\phi_A\nabla_{BA'}\phi^B = -\mu\phi_A\chi_{A'},$$

and if $C_a = \phi_A\chi_{A'}$ then equations (2.5.23–4) imply

$$\nabla_b F_a{}^b + M_a = -2\mu C_a, \qquad \nabla_B G_a{}^b + N_a = -2\mu C_a,$$

where G_{ab}, N_a are the obvious analogues of F_{ab}, M_a. Hence establish the tensorial equations governing Dirac's theory

$$F_{ab}\nabla_e F_c{}^e + F_{ae}\nabla_c F_b{}^e = -2\mu F_{ab}C_c,$$
$$G_{ab}\nabla_e G_c{}^e + G_{ae}\nabla_c G_b{}^e = -2\mu G_{ab}C_c,$$
$$C_a C_b = F_a{}^c G_{cb}.$$

2.6   The Newman-Penrose formalism

As was mentioned in section 2.5 many explicit calculations are most efficiently performed using a tetrad of vectors. While orthonormal tetrads had been used for some time, the idea of adapting tetrads of null vectors to spinors seems to have been developed first by Newman and Penrose, 1962. The idea is most easily understood as the translation of a spin basis (o, ι) to the real spacetime. As was shown in section 2.3, (o, ι) induce four null vectors, a **NP null tetrad** via

$$l^a = o^A\bar{o}^{A'}, \quad n^a = \iota^A\bar{\iota}^{A'}, \quad m^a = o^A\bar{\iota}^{A'}, \quad \bar{m}^a = \iota^A\bar{o}^{A'},$$
$$l_a = o_A\bar{o}_{A'}, \quad n_a = \iota_A\bar{\iota}_{A'}, \quad m_a = o_A\bar{\iota}_{A'}, \quad \bar{m}_a = \iota_A\bar{o}_{A'}. \qquad (2.6.1)$$

Obviously l and n are real while m and $\bar{m}$ are complex conjugates. The directional derivatives along these directions are denoted by the conventional symbols

$$D = l^a\nabla_a, \quad \Delta = n^a\nabla_a, \quad \delta = m^a\nabla_a, \quad \bar{\delta} = \bar{m}^a\nabla_a. \qquad (2.6.2)$$

Clearly ∇_a is a linear combination of these operators. Indeed using exercise 2.3.3

$$
\begin{aligned}
\nabla_a &= g_a{}^b \nabla_b \\
&= (n_a l^b + l_a n^b - \overline{m}_a m^b - m_a \overline{m}^b) \nabla_b \\
&= n_a D + l_a \Delta - \overline{m}_a \delta - m_a \overline{\delta}.
\end{aligned}
\tag{2.6.3}
$$

The idea now is to replace ∇_a by (2.6.3) and then to convert all remaining tensor equations to sets of scalar ones by contraction with the NP null tetrad. Although this may lead to a vast set of equations, they usually possess discrete symmetries, and because only scalars are involved they are usually easier to handle in specific calculations. As presented above the NP formalism is independent of spinors, and indeed there are a number of practitioners who claim to understand nothing of spinors! However the apparently fortuitous notation was not chosen at random, but has a spinorial origin, and with practice spinors are usually the most efficient tools for carrying out calculations.

We start by considering the connection. In a tetrad formalism this is described by the Ricci rotation coefficients (2.5.1). Each scalar coefficient is of the form

(basis vector).((directional derivative)(basis vector)),   (2.6.4)

and the obvious spinor analogue is given by (2.5.4). Each is of the form

$$
\alpha^A \nabla \beta_A,
$$

where α and β are (o, ι) and ∇ is one of $D, \Delta, \delta, \overline{\delta}$. Note that since $o_A \iota^A = 1$, $o^A \nabla \iota_A = \iota^A \nabla o_A$ and so there are three independent choices for α, β. Since there are four choices for ∇ there will be 12 complex rotation coefficients corresponding to the 24 real rotation coefficients. These are given in the box in appendix B. We could also derive them vectorially by noting that since $l^a l_a = 0$ it follows that $l^a \nabla l_a = 0$, so that there are only 6 choices of vectors in (2.6.4) and there are again 24 real coefficients. To see the relation between the two representations consider e.g.

$$\kappa = m^a D l_a$$
$$= o^A \bar{\iota}^{A'} D(o_A \bar{o}_{A'})$$
$$= o^A o_A \bar{\iota}^{A'} D\bar{o}_{A'} + o^A \bar{o}_{A'} \bar{\iota}^{A'} Do_A$$
$$= o^A Do_A.$$

Notice in particular that $\alpha, \beta, \gamma, \epsilon$ have particularly complicated vector descriptions.

We now derive the NP description of electromagnetism, and for simplicity we assume there are no sources. As we have seen the Maxwell field tensor can be described by a symmetric 2-spinor ϕ_{AB}. This generates 3 complex scalars via

$$\phi_0 = \phi_{AB} o^A o^B, \qquad \phi_1 = \phi_{AB} o^A \iota^B, \qquad \phi_2 = \phi_{AB} \iota^A \iota^B. \quad (2.6.5)$$

(ϕ_n contains n ι's and $2-n$ o's.) Recalling that $\epsilon^{AB} = o^A \iota^B - \iota^A o^B$, and that $\phi_{AB} = \epsilon_A{}^C \epsilon_B{}^D \phi_{CD}$ one immediately obtains

$$\phi_{AB} = \phi_2 o_A o_B - 2\phi_1 o_{(A} \iota_{B)} + \phi_0 \iota_A \iota_B. \quad (2.6.6)$$

It is not difficult to relate the ϕ_i to the components of **E**, **B** as defined by the orthonormal tetrad (2.3.3) induced by the NP null tetrad. For example

$$E_{\hat{3}} - iB_{\hat{3}} = F_{\hat{0}\hat{3}} + iF_{\hat{1}\hat{2}} = (F + iF^*)_{\hat{0}\hat{3}}$$
$$= \tfrac{1}{2}(F + iF^*)_{ab}(l + n)^a(l - n)^b$$
$$= (F + iF^*)_{ab} n^a l^b, \qquad \text{using the symmetries of } F,$$
$$= 2\phi_{AB} \epsilon_{A'B'} \iota^A \bar{\iota}^{A'} o^B \bar{o}^{B'} \qquad \text{using (2.4.2)},$$
$$= -2\phi_{AB} \iota^A o^B$$
$$= -2\phi_1.$$

(2.6.1) Exercise. Show that

$$E_{\hat{1}} - iB_{\hat{1}} = \phi_0 - \phi_2, \qquad E_{\hat{2}} - iB_{\hat{2}} = i(\phi_0 + \phi_2), \qquad E_{\hat{3}} - iB_{\hat{3}} = -2\phi_1.$$

The Maxwell equations are

$$0 = \nabla^{AA'} \phi_{AB} = \epsilon^{AC} \nabla_{AA'} \phi_{CB} = (o^A \iota^C - \iota^A o^C) \nabla_{AA'} \phi_{CB}. \quad (2.6.7)$$

There are four complex equations here given by $A', B = 0, 1$ corresponding to the eight real Maxwell equations. For example we may multiply (2.6.7) by $\bar{o}^{A'} o^B$ obtaining

$$\left(o^A \bar{o}^{A'} o^B \iota^C - \iota^A \bar{o}^{A'} o^B o^C\right) \nabla_{AA'} \phi_{BC} = 0,$$

or

$$\left(o^B \iota^C D - o^B o^C \bar{\delta}\right)\left[\phi_0 \iota_B \iota_C - \phi_1 o_B \iota_C - \phi_1 \iota_B o_C + \phi_2 o_B o_C\right] = 0,$$

or

$$- \phi_0 \iota^C D \iota_C + D\phi_1 + \phi_1 \iota^C D o_C - \phi_1 o^B D \iota_B + \phi_2 o^B D o_B - \bar{\delta}\phi_0 + \\ \phi_0 o^C \bar{\delta} \iota_C + \phi_0 o^B \bar{\delta} \iota_B - \phi_1 o^B \bar{\delta} o_B - \phi_1 o^C \bar{\delta} o_C = 0,$$

or

$$D\phi_1 - \bar{\delta}\phi_0 = (\pi - 2\alpha)\phi_0 + 2\rho\phi_1 - \kappa\phi_2.$$

A complete set of source-free Maxwell equations is given in appendix B.

We may decompose the trace-free Ricci tensor as

$$\begin{aligned}
\Phi_{00} &= \Phi_{ABA'B'} o^A o^B \bar{o}^{A'} \bar{o}^{B'}, \\
\Phi_{01} &= \Phi_{ABA'B'} o^A o^B \bar{o}^{A'} \bar{\iota}^{B'}, \\
\Phi_{02} &= \Phi_{ABA'B'} o^A o^B \bar{\iota}^{A'} \bar{\iota}^{B'}, \\
\Phi_{10} &= \Phi_{ABA'B'} o^A \iota^B \bar{o}^{A'} \bar{o}^{B'}, \\
\Phi_{11} &= \Phi_{ABA'B'} o^A \iota^B \bar{o}^{A'} \bar{\iota}^{B'}, \qquad\qquad (2.6.8) \\
\Phi_{12} &= \Phi_{ABA'B'} o^A \iota^B \bar{\iota}^{A'} \bar{\iota}^{B'}, \\
\Phi_{20} &= \Phi_{ABA'B'} \iota^A \iota^B \bar{o}^{A'} \bar{o}^{B'}, \\
\Phi_{21} &= \Phi_{ABA'B'} \iota^A \iota^B \bar{o}^{A'} \bar{\iota}^{B'}, \\
\Phi_{22} &= \Phi_{ABA'B'} \iota^A \iota^B \bar{\iota}^{A'} \bar{\iota}^{B'}.
\end{aligned}$$

Since $\Phi_{ij} = \overline{\Phi}_{ji}$ there are 9 independent real quantities, as might have been expected.

The ten independent components of the Weyl tensor may be combined into 5 complex scalars Ψ_n defined by

$$\begin{aligned}
\Psi_0 &= \Psi_{ABCD} o^A o^B o^C o^D, \\
\Psi_1 &= \Psi_{ABCD} o^A o^B o^C \iota^D, \\
\Psi_2 &= \Psi_{ABCD} o^A o^B \iota^C \iota^D, \qquad\qquad (2.6.9) \\
\Psi_3 &= \Psi_{ABCD} o^A \iota^B \iota^C \iota^D, \\
\Psi_4 &= \Psi_{ABCD} \iota^A \iota^B \iota^C \iota^D.
\end{aligned}$$

The equations defining the curvature tensor components in terms of derivatives and products of Ricci rotation coefficients, the so-called *field equations* together with the Bianchi identities are given in appendix B. Although they look fearfully complicated things are not as bad as they seem.

Since we are dealing only with scalars then for any NP quantity ϕ

$$\nabla_{[a}\nabla_{b]}\phi = 0. \tag{2.6.10}$$

Before we translate this into NP notation we note the following equations which follow from (2.6.3)

$$
\begin{aligned}
&Do_A = \epsilon o_A - \kappa \iota_A, &&D\iota_A = \pi o_A - \epsilon \iota_A, \\
&\Delta o_A = \gamma o_A - \tau \iota_A, &&\Delta \iota_A = \nu o_A - \gamma \iota_A, \\
&\delta o_A = \beta o_A - \sigma \iota_A, &&\delta \iota_A = \mu o_A - \beta \iota_A, \\
&\bar\delta o_A = \alpha o_A - \rho \iota_A, &&\bar\delta \iota_A = \lambda o_A - \alpha \iota_A.
\end{aligned}
\tag{2.6.11}
$$

For practical reference these equations are repeated in appendix B, together with the equivalent equations for the derivatives of the NP basis vectors.

Now reconsider (2.6.10) multiplied by say $l^a m^b$

$$m^b D \nabla_b \phi - l^a \delta \nabla_a \phi = 0,$$

or

$$D\delta\phi - (Dm^b)\nabla_b\phi - \delta D\phi + (\delta l^a)\nabla_a\phi = 0. \tag{2.6.12}$$

Now

$$
\begin{aligned}
Dm^b &= D(o^B \bar\iota^{B'}) \\
&= \epsilon o^B \bar\iota^{B'} - \kappa \iota^B \bar\iota^{B'} + \bar\pi o^B \bar o^{B'} - \bar\epsilon o^B \bar\iota^{B'} \\
&= \bar\pi l^b - \kappa n^b + (\epsilon - \bar\epsilon)m^b.
\end{aligned}
$$

Similarly

$$\delta l^a = (\bar\alpha + \beta)l^a - \bar\rho m^a - \sigma \overline{m}^a.$$

Thus (2.6.12) becomes

$$(D\delta - \delta D)\phi = (\bar\pi - \bar\alpha - \beta)D\phi - \kappa\Delta\phi + (\bar\rho + \epsilon - \bar\epsilon)\delta\phi + \sigma\bar\delta\phi. \tag{2.6.13}$$

The four commutator equations like this are given in appendix B.

Although the system of NP equations may seem, at first sight, to be horrendous, there are two discrete symmetries which can be used to simplify calculations. It is easy to see that the following two operations preserve spin bases

$$' : \qquad o^A \mapsto i\iota^A, \quad \iota^A \mapsto io^A, \quad \bar{o}^{A'} \mapsto -i\bar{\iota}^{A'}, \quad \bar{\iota}^{A'} \mapsto -i\bar{o}^{A'},$$

$$* : \qquad o^A \mapsto o^A, \quad \iota^A \mapsto \iota^A, \quad \bar{o}^{A'} \mapsto -\bar{\iota}^{A'}, \quad \bar{\iota}^{A'} \mapsto -\bar{o}^{A'}.$$

(2.6.2) Problem. Investigate what happens to the NP null tetrad, the connection coefficients and the Maxwell scalars ϕ_n. Hence show that the 4 Maxwell equations are just $'$, $*$ and $*'$ transforms of one equation. (Similar simplifications occur for the "field" equations and the Bianchi identities.)

A common practical problem is the computation of the NP scalars. Suppose a coordinate chart (x^a) and NP tetrad have been given, so that the components, e.g., $l^a(x)$, are known functions. Clearly it is an elementary exercise in algebra to compute both the covector components, e.g., $l_a(x)$ and the line element $g_{ab}(x)$ using exercises 2.3.1, 2.3.3. However since, e.g., $\kappa = m^a D l_a = m^a l^b \nabla_b l_a$ involves a covariant derivative it looks as though it is necessary to compute the Christoffel symbols from g_{ab}. Fortunately this is not the case.

Suppose first that all of the NP scalars are required. One uses the commutator equations, e.g., (2.6.13), listed in appendix B. For each **fixed** a one may regard x^a as a scalar field on the manifold and apply the commutator to it. To make this clear we attach a caret to the index. Then $\delta x^{\hat{a}} = m^{\hat{a}}$ is also a scalar field on the manifold. Thus (2.6.13) reads

$$Dm^{\hat{a}} - \delta l^{\hat{a}} = (\bar{\pi} - \bar{\alpha} - \beta)l^{\hat{a}} - \kappa n^{\hat{a}} + (\bar{\rho} + \epsilon - \bar{\epsilon})m^{\hat{a}} + \sigma \bar{m}^{\hat{a}}. \quad (2.6.14)$$

Since we are dealing only with scalar functions the remaining derivatives in (2.6.14) are partial ones, i.e., the connection is not needed. Applying the commutators to each of the fields $x^{\hat{a}}$ one obtains 24 real linear algebraic equations for the 12 complex connection coefficients, and these can be solved to obtain the NP scalars explicitly. The "field" equations can now be used to obtain the curvature components, using only partial differentiation and simple algebra.

If only a small subset of connection coefficients is needed there is
an alternative procedure which has been implicit in the literature,
but has only recently been described explicitly by Cocke, 1989.
Note that

$$\kappa = m^a D l_a = m^a D l_a - l^a \delta l_a$$
$$= 2 l^a m^b \nabla_{[a} m_{b]}, \qquad (2.6.15)$$

where the second equality follows since l is a null vector. Equation
(2.6.15) involves a "curl" of a vector and so only partial derivatives
are needed.

(2.6.3) Exercise. Obtain the expressions

$$\rho = \overline{m}^a m^b \nabla_{[a} l_{b]} - \overline{m}^a l^b \nabla_{[a} m_{b]} + l^a m^b \nabla_{[a} \overline{m}_{b]},$$

$$\beta = \tfrac{1}{2} \left(m^a n^b \nabla_{[a} l_{b]} - n^a l^b \nabla_{[a} m_{b]} + l^a m^b \nabla_{[a} n_{b]} \right) - m^a \overline{m}^b \nabla_{[a} m_{b]}.$$

The remaining expressions can be found using the $'$ and $*$
transforms.

In actual calculations use is often made of Lorentz transformations to simplify calculations. We need therefore to describe how
such transformations affect the NP coefficients. Let (o, ι) be a spin
basis. Then $(\tilde{o}, \tilde{\iota}) = (\lambda o, \lambda^{-1} \iota)$ is also a spin basis for any complex
λ. Setting $\lambda = a \exp(i\theta)$ we see that the NP tetrad transforms as

$$\tilde{l} = a^2 l, \quad \tilde{n} = a^{-2} n, \quad \tilde{m} = e^{2i\theta} m. \qquad (2.6.16)$$

Such transformations with $\theta = 0$ are called **boosts** and if $\theta \neq 0$,
$a = 1$ we have a **spin**. The general case is a combination of the
two. The behaviour of the NP scalars under boosts and spins is
given explicitly in appendix B.

A second subclass of spin basis transformations is given by
$(\hat{o}, \hat{\iota}) = (o, \iota + co)$, where c is an arbitrary complex number. The
NP tetrad transforms according to

$$\hat{l} = l, \quad \hat{n} = n + cm + \bar{c}\overline{m} + c\bar{c}l, \quad \hat{m} = m + \bar{c}l. \qquad (2.6.17)$$

(2.6.17) is called a **null rotation about l**. The behaviour of the
NP scalars under such rotations is given explicitly in appendix B.

By interchanging o and ι, the $'$ transform, one may also construct null rotations about n. These transformations depend on two real and two complex parameters, in total six real ones, and so constitute a representation of the Lorentz group.

2.7   Focussing and shearing

Perhaps the simplest application of the NP formalism is to the propagation of light rays.

We therefore consider a beam of light rays. The beam can be represented by a congruence of geodesics γ with tangent vector l^a and connecting vector z^a. The concept of "connecting vector" was fixed by definition 1.9.6, and it may prove helpful to review the discussion of geodesic deviation given in section 1.9. By definition their commutator vanishes

$$[l, z]^a = 0. \tag{2.7.1}$$

At some point p on each geodesic γ we choose a spinor o such that $o^A \bar{o}^{A'}$ is tangent to γ. We then complete o to a spin basis (o, ι) at p. We next propagate o and ι along γ by the conditions

$$Do^A = D\iota^A = 0. \tag{2.7.2}$$

This means that o and ι are parallelly propagated along γ and so remain a spin basis at each point of γ. Further identifying l^a with $o^A \bar{o}^{A'}$ it is clear that $Dl^a = Do^A \bar{o}^{A'} = 0$, so that γ is affinely parametrized.

Suppose that at p the connecting vector is orthogonal to γ i.e., $l^a z_a = 0$. Then

$$D(l^a z_a) = l^a Dz_a = l^a \nabla_l z_a = l^a \nabla_z l_a = \tfrac{1}{2} \nabla_z (l^a l_a) = 0,$$

where the third equality follows from (2.7.1). Thus z^a is always orthogonal to γ. We now construct the NP tetrad $(l, n, m, \overline{m})$ induced by the spin basis. Since z^a is real and orthogonal to l^a there exist a real u and complex z such that

$$\begin{aligned} z^a &= ul^a + \bar{z}m^a + z\overline{m}^a \\ &= uo^A \bar{o}^{A'} + \bar{z}o^A \iota^{A'} + z\bar{o}^{A'} \iota^A. \end{aligned} \tag{2.7.3}$$

We next compute

$$Dz^a = \nabla_l z^a = \nabla_z l^a = uDl^a + \bar{z}\delta l^a + z\bar{\delta}l^a,$$

or

$$o^A\bar{o}^{A'} Du + o^A\bar{\iota}^{A'} D\bar{z} + \iota^A\bar{o}^{A'} Dz$$
$$= \bar{z}o^A\delta\bar{o}^{A'} + \bar{z}\bar{o}^{A'}\delta o^A + zo^A\bar{\delta}\bar{o}^{A'} + z\bar{o}^{A'}\bar{\delta}o^A.$$

Multiplying by $o_A \bar{\iota}_{A'}$ gives

$$-Dz = -\bar{z}o_A\delta o^A - zo_A\bar{\delta}o^A,$$

i.e.,

$$Dz = -\rho z - \sigma\bar{z}. \tag{2.7.4}$$

The interpretation of z is as follows. Consider the projection of z^a onto the spacelike 2-plane spanned by $m^a, \overline{m}^a$, or equivalently $e_{\hat{1}}{}^a, e_{\hat{2}}{}^a$ from (2.3.3). Write the projection as

$$\sqrt{2}(xe_{\hat{1}} - ye_{\hat{2}}) = x(m + \overline{m}) + y(m - \overline{m})/i$$
$$= \bar{z}m + z\overline{m},$$

where $z = x + iy$, which is consistent with (2.7.3). Thus z describes the projection in an Argand 2-plane spanned by $m^a, \overline{m}^a$.

We now analyse equation (2.7.4). Suppose first that $\sigma = 0$ while ρ is real, i.e., $Dz = -\rho z$, or

$$Dx = -\rho x, \qquad Dy = -\rho y.$$

This is an isotropic magnification at a rate $-\rho$. Next suppose that $\sigma = 0$ while $\rho = -i\omega$, so that $Dz = i\omega z$, or

$$Dx = -\omega y, \qquad Dy = \omega x,$$

which corresponds to a rotation with angular velocity ω.

Next consider the case where $\rho = 0$ and σ is real . Then

$$Dx = -\sigma x, \qquad Dy = \sigma y,$$

which represents a volume-preserving shear at a rate σ with principal axes along the x and y axes.

(2.7.1) Exercise. Show that multiplying σ by $e^{i\alpha}$ rotates the principal shear axes by $\frac{1}{2}\alpha$.

Since (2.7.4) is a linear equation, the general case is a superposition of these effects. The congruence, or more precisely the projection of the connecting vectors onto an orthogonal spacelike 2-surface, is expanded, rotated and sheared. This could be realized experimentally as follows. Suppose an opaque screen with a small hole cut in it was placed orthogonal to a stationary beam of light rays, and a second screen was used to capture the image of the hole. By moving the second screen relative to the first the shape of the image would change because of these effects. This is an important feature in observational cosmology. One cannot draw conclusions about the intrinsic nature of an astronomical object from a telescope image without first disentangling these optical effects.

Since $Do^A = 0$ it follows immediately that $\kappa = \epsilon = 0$. Let us assume that l^a is orthogonal to a family of hypersurfaces, the *wave fronts*, i.e., there exist scalar fields v, w such that

$$l_a = v\nabla_a w. \tag{2.7.5}$$

Then it is easy to show that

$$l_{[a}\nabla_b l_{c]} = 0. \tag{2.7.6}$$

In fact the converse holds: (2.7.6) implies (2.7.5) as a consequence of Frobenius' theorem. Therefore a vector field l_a satisfying (2.7.6) is said to be *hypersurface-orthogonal*. Clearly hypersurface-orthogonality implies

$$m^a \overline{m}^b n^c l_{[a}\nabla_b l_{c]} = 0,$$

which implies that ρ is real. In this case the NP field equations (a), (b) of appendix B reduce to

$$D\rho = \rho^2 + \sigma\bar\sigma + \Phi_{00}, \qquad D\sigma = 2\rho\sigma + \Psi_0, \tag{2.7.7}$$

which imply

$$D^2 z = -z\Phi_{00} - \bar{z}\Psi_0. \qquad (2.7.8)$$

Now suppose that we have an initially parallel circular beam, i.e., $\rho = \sigma = 0$, in a region in which $\Phi_{00} = \Psi_0 = 0$. The beam enters a region where $\Phi_{00} \neq 0, \Psi_0 = 0$. It is clear that σ remains zero, but what happens to ρ? Now for all reasonable matter the energy density is positive, which we express by the **weak energy condition**:

$$T_{ab}t^a t^b \geq 0 \qquad \text{for all timelike } t^a, \qquad (2.7.9)$$

where T^{ab} is the energy-momentum tensor. (See e.g., Hawking and Ellis, 1973, for a detailed discussion of energy conditions.) By continuity this must hold for null vectors and so $T_{ab}l^a l^b \geq 0$. Now the Einstein field equations

$$T_{ab} = -(8\pi G)^{-1}(R_{ab} - \tfrac{1}{2}g_{ab}R)$$

imply $R_{ab}l^a l^b \leq 0$, and so $\Phi_{00} \geq 0$ for all physically reasonable matter. Φ_{00} represents the matter crossing the geodesics transversally. Now $D\rho = \rho^2 + \Phi_{00} \geq 0$ and so provided the strict inequality holds, ρ becomes positive. It is easy to see that $\rho \to \infty$ in a finite parameter time. For suppose v is an affine parameter and $\rho = \rho_o$ at $v = 0$. Consider first

$$\frac{d\rho}{dv} = D\rho = \rho^2.$$

The solution of this equation viz. $1/(\rho_0 - \rho)$ certainly becomes infinite when $v = 1/\rho_o$. But for the exact equation $D\rho$ must be even larger and so infinite growth cannot be avoided. When $\rho \to \infty$ we have a **caustic** which may also be a **curvature singularity**. (Geometrical singularities are discussed in some detail in Hawking and Ellis, 1973.) In optics this behaviour with $\sigma = 0$ is called **anastigmatic focussing**. Now suppose $\Psi_0 \neq 0$, which implies that $\sigma \neq 0$. Ψ_0 represents the gravitational radiation crossing the rays transversally. If shear is present then a caustic still occurs. For even if $\Phi_{00} = 0$, $D\rho = \rho^2 + \sigma\bar{\sigma}$ and the σ term generates focussing. This is an **astigmatic lens**.

(2.7.2) Exercise. Let $D = d/dv$ and solve (2.7.8) for $\Psi_0 = A\delta(v), \Phi_{00} = B\delta(v)$, where A, B are real constants, $B > 0$, and $\rho = \sigma = 0$ initially. (The general case is a "superposition" of such sources.)

We still have the problem of solving the NP equations. Although this is a formidable task, some sub-problems are relatively easy. Consider the "D-equations" which are the restriction of the NP equations to a null hypersurface with l^a tangent to the generators. We demonstrate that this reduces the NP equations to a linear system of ordinary differential equations. For simplicity we shall assume a vacuum and assume the tetrad has been chosen to set $\kappa = \epsilon = \pi = 0$. We also assume that Ψ_0 is given as free data. Define

$$X = \begin{pmatrix} \rho & \sigma \\ \bar{\sigma} & \rho \end{pmatrix} \qquad F = \begin{pmatrix} \Phi_{00} & \Psi_0 \\ \bar{\Psi}_0 & \Phi_{00} \end{pmatrix}$$

Then equations a), b) of appendix B imply

$$DX = X^2 + F.$$

We next define Y via $X = -(DY)Y^{-1}$. Like X and F, Y is Hermitian. This equation can also be written as $DY = -XY$. Now Y has to satisfy

$$D^2 Y = -FY.$$

Once this second order linear equation has been solved for Y we may make a non-linear transformation to recover X. Now equations c), d), e), Ba) of appendix B form a coupled linear system for the unknowns $\tau, \alpha, \beta, \Psi_1$. Once these have been solved the equations f), g), h), Bc) can be treated as a linear system for $\gamma, \lambda, \mu, \Psi_2$. Finally i), Be) form a linear system for ν, Ψ_3, and then Bg) is a linear equation for Ψ_4.

2.8   Goldberg Sachs et al.

Although the title sounds like the name of a stockbrokers' firm, it actually refers to a group of theorems. It is clear from the discussion in section 2.7 that the analysis of the field equations would be considerably simplified if σ vanished for a congruence of null geodesics. These theorems establish necessary and sufficient conditions for this to obtain. While these theorems appear to have no immediately utilizable consequences they serve a useful pedagogical service as an introduction to manipulating the NP equations. However as a precursor we consider the problem of *integrability conditions*. Suppose the scalar field $\phi(t, x)$ satisfies the equations

$$\frac{\partial \phi}{\partial t} = f(t, x), \qquad \frac{\partial \phi}{\partial x} = g(t, x), \qquad (2.8.1)$$

where f, g are known functions. Can a solution for ϕ be found? Locally this is a well-understood problem of the type $\vec{\nabla}\phi = \vec{F}$, and has a local solution if and only if $\vec{\nabla} \times \vec{F} = \vec{0}$, or

$$\frac{\partial f}{\partial x} = \frac{\partial g}{\partial t}. \qquad (2.8.2)$$

Next consider the generalization

$$A^a \nabla_a \phi = f, \qquad B^a \nabla_a \phi = g, \qquad (2.8.3)$$

where A^a, B^a are locally transverse vector fields which are *surface-forming*, i.e., there exist scalar fields α and β such that

$$[A, B] = \alpha A + \beta B. \qquad (2.8.4)$$

The reason for this name should be obvious from the geometrical interpretation of the commutator given in section 1.6.

(2.8.1) THEOREM

A necessary and sufficient condition for solutions of the system (2.8.3) to exist is

$$A^a \nabla_a g - B^a \nabla_a f = \alpha f + \beta g. \qquad (2.8.5)$$

Proof: As in chapter 1 we adopt the notation $Af = A^a \nabla_a f$. Now suppose first that (2.8.3) has a solution ϕ. Then (2.8.4) implies

$$A(B\phi) - B(A\phi) = \alpha A\phi + \beta B\phi.$$

Setting $A\phi = f$, $B\phi = g$ we obtain (2.8.5) immediately. Suppose conversely that (2.8.5) holds. We adopt the special coordinate system of lemma 1.6.2 to set $A = \partial/\partial y^1$. Then we may solve $A\phi = f$ via

$$\phi(y^1, y^\alpha) = \int_{\tilde{y}}^{y^1} f(s, y^\alpha)\, ds,$$

where $\alpha = 2, \ldots, n$. We denote this solution by $\phi = A^{-1}f$. Let $B\phi - g = h$. We may choose $\tilde{y} = \tilde{y}(y^\alpha)$ to set $h = 0$ on an initial surface $y^1 = const$. Now

$$
\begin{aligned}
Ah &= ABA^{-1}f - Ag \\
&= BAA^{-1}f + [A, B]A^{-1}f - Ag \\
&= Bf + (\alpha A + \beta B)A^{-1}f - Ag \qquad \text{by (2.8.4)}, \\
&= \beta B\phi - \beta g \qquad \text{by (2.8.5)}, \\
&= \beta h.
\end{aligned}
$$

Since $h = 0$ initially we see that $h = 0$ and so $B\phi = g$, i.e., ϕ solves (2.8.3).

We start with two results concerning solutions of Maxwell's equations and shearfree congruences.

(2.8.2) THE MARIOT-ROBINSON THEOREM

Suppose that F_{ab} is a null electromagnetic field. Then the repeated principal null direction generates a geodesic shearfree null congruence.

Proof: Let the repeated principal spinor be o_A and set $\phi_{AB} = \phi o_A o_B$. Complete o to a spin basis (o, ι). Then from (2.6.5) it follows that $\phi_0 = \phi_1 = 0$, $\phi_2 = \phi$. Now the Maxwell equations

(Ma, Mc) of appendix B imply that $\sigma = \kappa = 0$. Further from (2.6.11) we have $Do_A = \epsilon o_A$. It follows that

$$Dl^a = D(o^a \bar{o}^{A'}) = (\epsilon + \bar{\epsilon})o^A \bar{o}^{A'} = (\epsilon + \bar{\epsilon})l^a,$$

so that the congruence with tangent vector l^a is both shearfree and geodesic, (although it need not be affinely parametrized.)

There is a converse result.

(2.8.3) ROBINSON'S THEOREM

Suppose spacetime contains a geodesic shearfree null congruence. Then there exists a null electromagnetic field, i.e., a solution of the Maxwell equations with repeated principal null direction tangent to the congruence.

Proof: Just as in the construction described in section 2.7 we may choose a spin basis (o, ι) with $o^A \bar{o}^{A'}$ tangent to the congruence, and o^A, ι^A parallelly propagated along the congruence, i.e., $Do^A = D\iota^A = 0$. Then equations (2.6.11) imply that $\kappa = \epsilon = \pi = 0$. The Maxwell equations for a null field $\phi_{AB} = \phi o_A o_B$ reduce to (Mb, Md) viz.

$$D\phi = \rho\phi, \qquad \delta\phi = (\tau - 2\beta)\phi. \qquad (2.8.6)$$

In general the equations (2.8.6) have no solution. However $[D, \delta] = -(\bar{\alpha} + \beta)D + \bar{\rho}\delta$, and so l^a and m^a are surface-forming. If we introduce $\theta = \ln\phi$ as a new dependent variable then equations (2.8.6) become

$$D\theta = \rho, \qquad \delta\theta = \tau - 2\beta.$$

The integrability condition becomes

$$D[\tau - 2\beta] - \delta\rho = -(\bar{\alpha} + \beta)\rho + \bar{\rho}(\tau - 2\beta). \qquad (2.8.7)$$

Using the field equations c), e), k) of appendix B this condition is easily seen to be satisfied identically. Thus the integrability condition for the Maxwell equations (2.8.6) is always satisfied at a point if the Maxwell equations hold at the point and the field equations (c,e,k) hold there. It is therefore possible to integrate the equations obtaining a solution satisfying the requirements of the theorem.

From exercise 2.4.4 and theorem 2.8.2 it is clear that plane electromagnetic waves can only propagate in spacetimes which admit a geodesic shearfree null congruence. We must therefore ask what restriction the latter condition imposes on a spacetime. The necessary and sufficient conditions are given by the following theorem.

(2.8.4) THE GOLDBERG-SACHS THEOREM

In a vacuum spacetime any two of the following conditions imply the third:

A) *The Weyl spinor* Ψ_{ABCD} *is algebraically special with n-fold repeated principal spinor* o, $(n = 2, 3, 4)$,
B) *either spacetime is flat or* o *generates a geodesic shearfree congruence,*
C) $\nabla^{AA'}\Psi_{ABCD}$ *contracted with* $(5 - n)$ o*'s vanishes.*

Proof: Assume A with $n = 2$, $\Psi_{ABCD} = o_{(A}o_B\alpha_C\beta_{D)}$. Clearly

$$o^A o^B o^C \Psi_{ABCD} = 0,$$

and so

$$o^A o^B o^C \nabla^{DD'}\Psi_{ABCD} + 3\Psi_{ABCD}o^A o^B \nabla^{DD'}o^C = 0. \qquad (2.8.8)$$

Now suppose that in addition B holds. If spacetime is flat then condition C obviously holds. If instead o is geodesic and shearfree then $o^C\delta o_C = o^C Do_C = 0$ and $o_C o_D \nabla^{DD'}o^C = 0$. It follows that the last term in (2.8.8) vanishes and so C is true. Suppose instead that C holds. Then

$$\Psi_{ABCD}o^A o^B \nabla^{DD'}o^C = 0,$$

and so $o_C o_D \nabla^{DD'} o^C = 0$ which proves B. The proof that B and C imply A is slightly messy. See e.g., section 7.3 of Penrose and Rindler, 1986.

(2.8.5) Exercise. Show that A and B imply C, and A and C imply B for $n = 3, 4$.

2.9 Plane and plane-fronted waves

In elementary courses on electromagnetism "plane waves" are defined to be solutions of Maxwell's equations in Minkowski space-time with $e^{ik_a x^a}$ dependence, where k_a is a constant null covector. Because of the linearity the Fourier decomposition is not essential, and so we consider a (slight) generalization.

(2.9.1) DEFINITION

*A **plane-fronted** electromagnetic wave in Minkowski spacetime is a solution of Maxwell's equations possessing a 1-parameter family of null plane wavefronts, $u = const.$ where*

$$u = -k_a(u)x^a, \qquad k_a \ null \ and \ nonzero. \qquad (2.9.1)$$

*In the special case where k_a is constant we have a **plane wave**.*

Note that on a surface $u = const.$ k_a is constant and so the wave fronts are still plane. A stationary torch should emit plane waves, but if it is waved around the waves should be plane-fronted. It is clear that definition 2.9.1 is highly non-covariant. If we want to study wave propagation in a general spacetime then we would expect 'plane-fronted' waves to be the simplest, and so we want a covariant definition of 'plane-fronted'. Such a definition is suggested by the following lemma.

(2.9.2) LEMMA

Consider a plane-fronted wave in Minkowski spacetime, and define
$l_a = \nabla_a u$, *where* u *is defined as in definition 2.9.1. Then* l_a *is null,*
geodesic, and free of rotation, divergence and shear.

Proof: It follows from (2.9.1) that

$$l_a = \nabla_a u = -k_a(u) - x^b k_b'(u)\nabla_a u, \qquad (2.9.2)$$

or

$$k_a = \lambda l_a, \qquad \lambda = -(1 + k_c'(u)x^c). \qquad (2.9.3)$$

Since k_a is null, so is l_a. Since l_a is a gradient $\nabla_{[a}l_{b]} = 0$, and so

$$Dl_a = l^b \nabla_b l_a = l^b \nabla_a l_b = \nabla(\tfrac{1}{2}l_b l^b) = 0,$$

and l_a is geodesic. Now differentiating (2.9.3) gives

$$k_a'(u)l_b = \lambda \nabla_b l_a + l_a \nabla_b \lambda. \qquad (2.9.4)$$

Let us complete l^a to a NP tetrad. Then transvecting (2.9.4)
with m^b shows that $\lambda \delta l_a + l_a \delta \lambda = 0$ and $\lambda m^a \delta l_a = 0$ or $\sigma = 0$.
Similarly $\rho = m^a \bar{\delta} l_a = 0$. Note further that the special case
$k_a = const.$ implies $\lambda = -1$, and so from (2.9.4) we may conclude
that $\nabla_a l_b = 0$.

Thus we are led to the following coordinate-free definition.

(2.9.3) DEFINITION

Consider a wave with wavefronts $u = const$. *Let* $l_a = \nabla_a u$. *If* l *is*
null, geodesic, and free of rotation, shear and divergence then the
wave is **plane-fronted**. *If in addition* $\nabla l_b = 0$ *the wave is said to*
be **plane parallel** *or* **pp**. *(In this case, as we shall see in theorem*
2.9.4 the nullity condition can be dropped.)

In the electromagnetic case Robinson's theorem guarantees the existence of plane-fronted and pp waves, and they are all of type N. We turn therefore to the gravitational case. It follows from the Goldberg-Sachs theorem that plane-fronted waves can only occur in an algebraically special spacetime. Suppose that we have a plane-fronted wave. We may without loss of generality assume l is affinely parametrized with $\epsilon = 0$. This case is still too general for a complete analysis and so we investigate pp waves.

We now state and prove an exact result on pp gravitational radiation. Although the proof is long it is worthwhile working through it. This is a classic example of the mixed use of tensorial, spinorial and NP techniques, and the complete discussion is not otherwise available.

(2.9.4) THEOREM

*Suppose a vacuum spacetime contains a non-zero covariantly constant or **parallel** vector field l*

$$\nabla l_b = 0. \qquad (2.9.5)$$

Then:

1) l is null, and the spacetime is algebraically special of type N with repeated principal null direction l,

2) coordinates (u, v, x, y) can be found so that the line element is

$$ds^2 = 2H(x^a)du^2 + 2dudv - dx^2 - dy^2, \qquad (2.9.6)$$

3) $H(x^a) = \Re(f(u, z))$ where $z = x + iy$ and f is analytic in z but otherwise arbitrary.

Proof: We start with part 1. Equation (2.9.5) implies $\nabla_{[a}\nabla_{b]}l_{CC'} = 0$, and the vacuum condition then implies

$$\Psi_{ABCD}l^{DD'} = 0. \qquad (2.9.7)$$

Now if $l^{DD'} \neq 0$ there must exist a β_D such that $\hat{o}^D = l^{DD'}\bar{\beta}_{D'} \neq 0$. Then equation (2.9.7) implies

$$\Psi_{ABCD}\hat{o}^D = 0. \qquad (2.9.8)$$

This means that the Weyl tensor is of type N with repeated principal spinor $\hat{o}$ i.e.

$$\Psi_{ABCD} = \Psi_4 \hat{o}^A \hat{o}^B \hat{o}^C \hat{o}^D.$$

(This can be seen by completing $\hat{o}$ to a spin basis and considering the Weyl equivalent of equation (2.6.6).) Now from (2.9.7) it follows that $\hat{o}_D l^{DD'} = 0$ from which we may deduce that there exists a γ such that $l_{DD'} = o_D \bar{\gamma}_{D'}$. The reality of l implies that $l_{DD'}$ must be Hermitian, i.e., $l_{DD'} = R\hat{o}_D \bar{\hat{o}}_{D'}$ where R is real. Thus l is null. By incorporating a factor $\sqrt{R}$ into $\hat{o}$ we obtain

$$l_{DD'} = \hat{o}_D \bar{\hat{o}}_{D'}. \tag{2.9.9}$$

To prove part 2 we first note that equations (2.9.5, 2.9.9) can be rewritten as

$$\nabla_{AA'}(\hat{o}_B \bar{\hat{o}}_{B'}) = 0, \tag{2.9.10}$$

which implies $\hat{o}^B \nabla_{AA'} \hat{o}_B = 0$ or

$$\nabla_{AA'} \hat{o}_B = i\alpha_{AA'} \hat{o}_B, \tag{2.9.11}$$

for some $\alpha_{AA'}$, and equation (2.9.10) implies that $\alpha_{AA'}$ is Hermitian. Further, in vacuum, equation (2.9.8) and the Ricci identity (exercise (2.5.5)) imply

$$0 = \Psi_{ABCD} o^D = \square_{AB} \hat{o}_C = \nabla_{A'(A} \nabla_{B)}{}^{A'} \hat{o}_C,$$

or, using (2.9.11)

$$0 = \hat{o}_C \nabla_{(A}{}^{A'} \alpha_{B)A'} + i\alpha_{A'(A} \alpha_{B)}{}^{A'} \hat{o}_C.$$

On examination the last term on the right hand side vanishes and the vanishing of the first implies that $\nabla_{[a} \alpha_{b]} = 0$, so that locally $\alpha_a = \nabla_a \alpha$ for some scalar field α. If we set $o_A = e^{-i\alpha} \hat{o}_A$ then (2.9.11) implies

$$\nabla_a o_A = 0. \tag{2.9.12}$$

Now (2.9.5) implies $\nabla_{[a} l_{b]} = 0$ and so locally $l_a = \nabla_a u$ for some scalar field u. We may choose a coordinate system (x^a) with $x^0 = u$ so that

$$l_a = \delta_a^0. \tag{2.9.13}$$

Next complete o to a spin basis (o, ι) and define

$$\omega_{ab} = 2l_{[a}m_{b]} = o_A o_B \epsilon_{A'B'}.$$

Clearly (2.9.12) implies $\nabla_a \omega_{bc} = 0$ or

$$(\nabla_a m_{[b})l_{c]} = 0. \tag{2.9.14}$$

This implies

$$\epsilon^{abcd} l_a \nabla_b m_c = 0, \tag{2.9.15}$$

or, using (2.9.13)

$$\epsilon^{\alpha\beta\gamma} \nabla_\beta m_\gamma = 0,$$

where greek indices range over 1 to 3. Thus there exists a scalar field $z = z(x^\gamma)$ such that locally

$$m_a = \nabla_a z + f(x^c)l_a. \tag{2.9.16}$$

(Note that (2.9.15) says nothing about m_0.) However since we have not fixed precisely the spin basis we have the freedom to add a multiple of o to ι, and it is clear that this can be used to set $f = 0$ in (2.9.16). Now (2.9.14) and the fact that $\nabla_a m_b$ is symmetric imply

$$\nabla_a m_b = g l_a l_b$$

for some scalar function g. Thus

$$m^a \nabla_a m_b = \overline{m}^a \nabla_a m_b = 0. \tag{2.9.17}$$

We now set $z = 2^{-1/2}(x + iy)$ and $X_a = \nabla_a x$, $Y_a = \nabla_a y$. Clearly $X_a X^a = Y_a Y^a = -1$, $X_a Y^a = 0$ so that u, x, y can be taken as independent coordinates. Further (2.9.17) implies

$$\nabla_X X = \nabla_X Y = \nabla_Y X = \nabla_Y Y = 0$$

so that x, y span a Euclidean 2-space. (The connection components vanish identically.) This 2-space will be called a *wave-surface*.

We now choose a wave surface in each $u = const.$ hypersurface such that the family forms a 3-surface (with coordinates u, x, y).

We define a scalar field v so as to vanish on this hypersurface, and to be propagated by

$$Dv = l^a \nabla_a v = 1. \qquad (2.9.18)$$

Setting $V_a = \nabla_a v$ we have on the initial hypersurface

$$m^a V_a = 0. \qquad (2.9.19)$$

Now

$$
\begin{aligned}
D(m^a V_a) &= (Dm_a)V^a + (\nabla_b V_a)m^a l^b \\
&= 0 + (\nabla_a V_b)m^a l^b \quad \text{using (2.9.14) and } \nabla_{[a} V_{b]} = 0, \\
&= \nabla_a (V_b l^b) m^a \\
&= \nabla_a (1) m^a = 0.
\end{aligned}
$$

Thus (2.9.19) holds everywhere. We shall use u, v, x, y as coordinates. It is clear that

$$l_a = (1, 0, 0, 0), \qquad m_a = 2^{-1/2}(0, 0, 1, i).$$

If we set

$$n_a = Hl_a + Jv_{,a} + Cm_a + \bar{C}\overline{m}_a,$$

then (2.9.18) implies that $J = 1$, while (2.9.19) implies that $C = 0$. Thus

$$n_a = (H, 1, 0, 0),$$

where H is to be determined. The line element (2.9.6) now follows from $g_{ab} = 2l_{(a}n_{b)} - 2m_{(a}\overline{m}_{b)}$.

We end by proving part 3. It is easy both to obtain the inverse of the metric tensor from (2.9.6) and derive

$$l^a = (0, 1, 0, 0), \quad n^a = (1, -H, 0, 0), \quad m^a = -2^{-1/2}(0, 0, 1, i). \qquad (2.9.20)$$

Thus $D = \partial/\partial v$ etc. We may now apply the commutators given in appendix B to the scalar fields u, v, x, y in turn. For example (2.9.20) implies

$$Du = 0, \qquad \Delta u = 1, \qquad \delta u = \bar{\delta}u = 0,$$

and the commutator equations applied to u give

$$\epsilon + \bar{\epsilon} = \kappa = 0, \qquad \tau = \bar{\alpha} + \beta, \qquad \rho = \bar{\rho}.$$

After constructing all four sets of four equations like this, and recalling that $\nabla l_b = 0$, we find that all of the connection scalars except ν vanish, and

$$DH = 0, \qquad \delta H = -\bar{\nu}. \tag{2.9.21}$$

Thus H is v-independent. All of the field equations of appendix B are trivially satisfied, except for (j) and (n). (n) asserts $\delta\nu = 0$ in a vacuum, i.e., $\delta\bar{\delta}H = 0$. Since the wave surface is flat this is equivalent to

$$H_{,xx} + H_{,yy} = 0, \tag{2.9.22}$$

which is sufficient to prove assertion 3).

The final equation (j) gives

$$\Psi_4 = -\bar{\delta}\bar{\delta}H = -\frac{\partial^2 H}{\partial y^2} + i\frac{\partial^2 H}{\partial x \partial y}. \tag{2.9.23}$$

(2.9.5) Exercise. Derive equations (2.9.21–23).

We may write

$$\Psi_4 = A e^{i\theta},$$

where $A(u, z)$ and $\theta(u, z)$ are real. If the phase θ of the wave is constant the wave is said to be ***linearly polarized***, while if A is constant on each wavefront $u = const$ the wave is called ***plane***. If A vanishes for $u < u_1$ and for $u > u_2$ the wave is said to be ***sandwich***.

Equation (2.9.6) represents a remarkable line element. We shall describe only one of its features. A geodesic will be said to be ***type 1*** if it lies in the 2-surface $u = const.$, $v = const.$, and ***type 2*** if it lies in the 2-surface $z = const.$ Type 1 geodesics reside in R^2 and so are ***complete***, i.e., infinitely extendable in both directions. Type 2 geodesics reside in a 2-surface with line element

$$ds^2 = 2H(u, z)du^2 + 2dudv.$$

By making a coordinate transformation $U = u$, $V = v + \int H(u,z)\,du$, with constant z we see immediately that this is 2-dimensional Minkowski space, and so type 2 geodesics are complete.

Note further that there is some coordinate gauge freedom. l_a is defined up to a constant factor, and so u is defined up to a constant linear transformation $u \mapsto \hat{u} = cu + d$. m^a is fixed up to a rotation $m^a e^{i\theta}$ and the addition of a multiple of l^a, and this gives $z \mapsto \hat{z} = ze^{i\theta} + p(u)$. The consequent change in v to preserve the line element is

$$v \mapsto \hat{v} = c^{-1}\left[v - \Re[ze^{i\theta}p'(u)] + q(u)\right],$$

where $q(u)$ is real. It is now straightforward to show that by means of a gauge transformation every geodesic can be converted to type 1 or 2. Thus the spacetime is geodesically complete.

(2.9.6) Problem. The discussion given here is based in part on Pirani, 1965, and Ehlers and Kundt, 1962. An alternative discussion of plane linearly polarized waves is given in section 35.9 of Misner, Thorne and Wheeler, 1973. Examine and compare these various approaches.

(2.9.7) Problem. Consider the pp wave vacuum spacetime with line element

$$ds^2 = 2H(u,x,y)du^2 + 2dudv - dx^2 - dy^2$$

of this section. Is the choice $H = \frac{1}{2}f(u)(x^2 - y^2)$ consistent with the vacuum condition? Make the coordinate transformation $(u,v,x,y) \to (U,V,X,Y)$ with

$$u = U, x = a(u)X, \qquad y = b(u)Y, \qquad v = V + \tfrac{1}{2}(X^2 aa' + Y^2 bb')$$

with $-a'/a = b'/b = f$. Suppose the metric represents a **weak sandwich wave**, i.e., f vanishes and spacetime is flat for $u < u_0$ and for $u > u_1$. Determine the linearized solution of the field equations. What happens to test particles as the wave passes through them? What condition has to be imposed on f to ensure that the wave is sandwich?

3

ASYMPTOPIA

3.1 Introduction

The title of this chapter deserves some explanation. Far from an isolated gravitational source one might expect an idealized simple description of spacetime — asymptotic Utopia. However the remark of the British Prime Minister Chamberlain in 1939 on Czechoslovakia is relevant here: "This is a far far away country about which we know very little."

Just as in special relativity, integral conservation laws can be derived in the standard way from vector divergence equations e.g.

$$\nabla_a J^a = 0. \tag{3.1.1}$$

However energy conservation is expressed locally by the vanishing of the divergence of the energy-momentum tensor T^{ab}

$$\nabla_b T^{ab} = 0, \tag{3.1.2}$$

and such an equation cannot lead to an integral conservation law. (The integrand would be vector-valued and one cannot add vectors at different points.) The situation is ameliorated if the spacetime possesses symmetries. For suppose K^b is a Killing vector. Then the current J^a defined by

$$J^a = T^{ab} K_b, \tag{3.1.3}$$

satisfies (3.1.1). For example the 4 translational Killing vectors in Minkowski spacetime lead to conservation of energy-momentum. In general however spacetime possesses no Killing vectors. For an isolated gravitating source one might expect spacetime to become

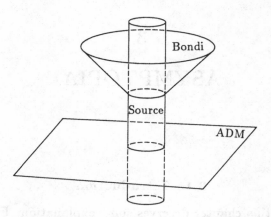

Fig. 3.1.1   The hypersurfaces on which the ADM and Bondi masses are measured.

flat asymptotically as one moved further away from it. By "going to infinity" one might acquire the Killing vectors necessary for integral conservation laws. Another viewpoint lends credence to this idea. A "gravitational energy density" would presumably be a quadratic function of the connection coefficients. Since these can be made to vanish locally such a density could not be localizable, but its integral over a large region might be well defined.

The investigation of asymptopia depends on how we approach it. Proceeding along spacelike hypersurfaces does not produce very interesting results. Imagine an isolated source initially at rest which suddenly emits a burst of gravitational radiation. Consider first a spacelike hypersurface which intersects the source during its quiescent state, and which leads to a "mass measured at infinity", the *ADM mass*, of M. Suppose a later spacelike hypersurface intersects the source during or after its active phase. One might expect the mass of the source to have decreased because of the energy emission. All of the radiation will however intersect this hypersurface, and allowing for its contribution one might expect the later ADM mass to be the same as the earlier one, which is indeed the case. If we want to measure the mass loss of the source it is clear that we should bend up the later hypersurface so that the already emitted radiation remains in its past, never intersecting it. If we bend the surface up until it becomes timelike then radiation emitted to the future of the intersection of the hypersurface with the source will intersect the surface a second time, and so will contribute twice to the mass! It seems clear therefore that we should bend the hypersurface such that it becomes null at far distances from the source, i.e., we should approach asymptopia along null directions. The corresponding mass is the **Bondi mass** and one might reasonably expect it to decrease for radiating sources.

Another major problem is that we have no exact radiating solutions which become flat asymptotically. Thus we can only guess what structure to expect far from an isolated source. This requires great care. If we set up axioms which are too strong we may find that only flat spacetime satisfies them! On the other hand if they are too weak we may be unable to draw useful conclusions. Perhaps the best way to get a feel for what is involved is by constructing asymptopia for Minkowski spacetime. While it is certainly possible to discuss asymptopia by taking appropriate limits as "$r \to \infty$" the analysis has to be done rather carefully, as we shall see in section 3.10. There is however a useful technical trick due to Penrose for avoiding such limits. One performs an isotropic scaling of (proper) distances with the scaling decreasing to zero as distance from the source increases; the total scaled distance to asymptopia then remains finite. Technically this is done by a conformal transformation. By convention we denote the original metric, *the physical metric* by $\tilde{g}$, and consider the transformation to *the unphysical metric* g given by

$$\tilde{g}_{ab} \mapsto g_{ab} = \Omega^2 \tilde{g}_{ab}, \qquad (3.1.4)$$

where $\Omega \geq 0$ is a smooth function which tends to zero far from the source. The adjective "conformal" is appropriate here since angles are preserved under (3.1.4). Unfortunately there is another convention in common usage in which the physical metric is g and the unphysical metric is $\hat{g}$, i.e., (3.1.4) becomes

$$g_{ab} \mapsto \hat{g}_{ab} = \Omega^2 g_{ab}. \qquad (3.1.5)$$

We shall use this convention in sections 3.10 and 3.11 to minimise the number of characters in the equations.

3.2   Asymtopia for Minkowski spacetime

We introduce standard Cartesian coordinates x^a and the standard Minkowski metric $\tilde{\eta}_{ab} = \mathrm{diag}(1, -1, -1, -1)$. We shall examine some examples of conformally related metrics

$$g_{ab} = \Omega_M^2 \tilde{\eta}_{ab}, \qquad (3.2.1)$$

where Ω_M is smooth and non-zero.

As a first example we investigate

$$\Omega_M = 1/X^2, \quad \text{where} \quad X^2 = \tilde{\eta}_{ab} x^a x^b, \qquad (3.2.2)$$

on the complement of the light cone of the origin. The first question is whether we can identify the "unphysical" spacetime with metric g given by (3.2.1). This turns out to be particularly easy if we introduce new coordinates y^a via the equations

$$y^a = -x^a/X^2, \quad x^a = -y^a/Y^2, \quad Y^2 = \tilde{\eta}_{ab} y^a y^b, \qquad (3.2.3)$$

so that

$$dy^a = -X^{-2} \left[\delta_c{}^a - 2X^{-2} x^a \tilde{\eta}_{cb} x^b \right] dx^c.$$

Now a simple calculation gives

$$\tilde{\eta}_{ab} dy^a dy^b = X^{-4} \tilde{\eta}_{ab}\, dx^a dx^b,$$

so that

$$\tilde{\eta}_{ab} dy^a dy^b = \Omega_M^2 \tilde{\eta}_{ab}\, dx^a dx^b. \qquad (3.2.4)$$

Thus as seen in the y-coordinates the "unphysical" spacetime is flat. It is clear too that in the complement of the light cone of the x-origin, curves tending to x-infinity, i.e., $X^2 \to \pm\infty$ (where the sign is positive for curves within the light cone and negative otherwise) approach the y-origin.

Consider first timelike geodesics through the x-origin

$$x^a = \lambda T^a, \tag{3.2.5}$$

where $\tilde{\eta}_{ab} T^a T^b = 1$, $T^0 > 0$, and $\lambda > 0$. It is clear that $X^2 = \lambda^2$. We may introduce a new parameter $\mu = -1/\lambda$, where the sign is chosen to ensure $d\mu/d\lambda > 0$. In terms of y^a the curve becomes

$$y^a = \mu T^a.$$

Note that, setting $\dot{y} = dy/d\mu$ etc.

$$\dot{y}^a = T^a, \qquad \ddot{y}^a = 0. \tag{3.2.6}$$

Thus a future-directed timelike geodesic through the x-origin approaching x-infinity maps to a future directed timelike geodesic approaching the y-origin (from below).

Consider next a timelike geodesic not passing through the x-origin

$$x^a = \lambda T^a + C^a,$$

where λ, T^a are as before, and C^a is spacelike and orthogonal to T^a. By means of a Lorentz transformation we can arrange that

$$T^a = (1, 0, 0, 0), \quad C^a = (0, C, 0, 0).$$

It is easy to see that $X^2 = \lambda^2 - C^2$ and defining μ as before, we find

$$y^a = \left(\frac{\mu}{1 - C^2 \mu^2}, \frac{-C\mu^2}{1 - C^2 \mu^2}, 0, 0 \right).$$

This curve also approaches the y-origin from below. However it is not a geodesic, for

$$\dot{y}^a(0) = T^a, \qquad \ddot{y}^a(0) = -2C^a.$$

Thus all future-pointing timelike geodesics approach the y-origin from below. All geodesics with the same x-velocity are tangent to each other at the y-origin, i.e., they have the same y-velocity there. However two parallel geodesics will have a relative y-acceleration at the y-origin equal to the spacelike x-separation.

(3.2.1) Exercise. **Perform the same analysis for spacelike geodesics. By considering null geodesics as limiting cases of both families of timelike geodesics and families of spacelike geodesics, show that a future-directed generator of the future null cone of the x-origin becomes a future-directed generator of the past null cone of the y-origin. How do two parallel null geodesics differ asymptotically?**

In practice (3.2.2) is a rather inconvenient choice of conformal factor. Consider physical Minkowski spacetime in spherical polar coordinates

$$d\tilde{s}^2 = dt^2 - dr^2 - r^2\, d\Sigma^2, \tag{3.2.7}$$

where

$$d\Sigma^2 = d\theta^2 + \sin^2\theta\, d\phi^2.$$

We now introduce the standard null coordinates

$$u = t - r, \qquad v = t + r, \qquad v \ge u, \tag{3.2.8}$$

so that

$$d\tilde{s}^2 = du\,dv - \tfrac{1}{4}(v - u)^2\, d\Sigma^2. \tag{3.2.9}$$

Now consider

$$ds^2 = \Omega_E^2\, d\tilde{s}^2, \tag{3.2.10}$$

with

$$\Omega_E^2 = \frac{4}{(1 + u^2)(1 + v^2)}, \tag{3.2.11}$$

i.e.

$$ds^2 = \frac{4}{(1 + u^2)(1 + v^2)}\, du\,dv - \frac{(u - v)^2}{(1 + u^2)(1 + v^2)}\, d\Sigma^2. \tag{3.2.12}$$

To interpret this metric it is convenient to introduce new coordinates

$$p = \arctan u, \qquad q = \arctan v, \qquad -\tfrac{1}{2}\pi < p \le q < \tfrac{1}{2}\pi. \tag{3.2.13}$$

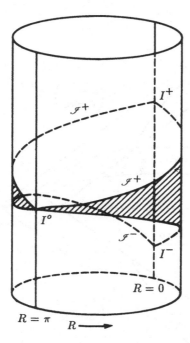

Fig. 3.2.1 The Einstein static universe. The coordinates θ, ϕ have been suppressed, and each point represents a 2-sphere of radius $\sin R$. The non-vertical lines represent the conformal boundaries of Minkowski spacetime and the interior is the Einstein static spacetime.

(3.2.2) Exercise. Show that

$$ds^2 = 4\,dp\,dq - \sin^2(p - q)\,d\Sigma^2, \qquad (3.2.14)$$

and if $T = p + q$, $R = q - p$, where $-\pi < T < \pi$, $-\pi < T - R < \pi$, $0 < R < \pi$

$$ds^2 = dT^2 - dR^2 - \sin^2 R\,d\Sigma^2. \qquad (3.2.15)$$

The line element (3.2.15) represents the product of the real line with a 3-space of constant curvature $+1$, the **Einstein static universe**. Suppressing the θ, ϕ-coordinates we can sketch it on a cylinder, fig. 3.2.1.

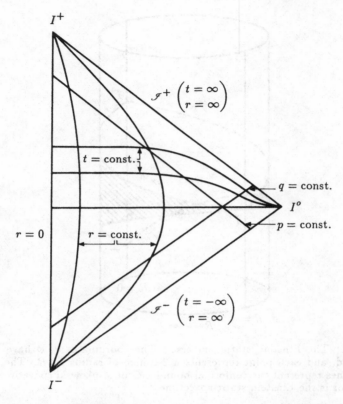

Fig. 3.2.2 A Penrose diagram for Minkowski spacetime. The angular co-ordinates have been suppressed, and lines of constant t, r, p and q are shown.

Next we shall consider a future-pointing timelike geodesic in the physical spacetime. Clearly $t \pm r \to \infty$, i.e., $u, v \to \infty$ or $p, q \to \frac{1}{2}\pi$, or $T \to \pi$, $R \to 0$. Thus all future-pointing timelike geodesics reach $T = \pi$, $R = 0$, which will be called I^+ or **future timelike infinity**. Note also that $t^2 - r^2 = uv$. Therefore near I^+

$$\Omega_E \sim 2/(t^2 - r^2) \sim 2\Omega_M.$$

It follows that apart from a trivial factor the two Ω's are asymptotically the same. Thus the asymptotic behaviour of parallel timelike geodesics is as stated previously on p.117. Similarly

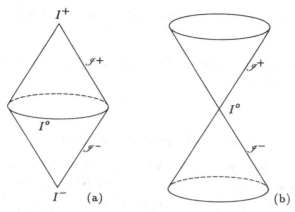

Fig. 3.2.3 A misleading version of the Penrose diagram for Minkowski spacetime (a), and the correct structure of spacelike infinity (b).

past-directed timelike geodesics approach ***past timelike infinity***, I^-, where $T = -\pi$, $R = 0$. Spacelike geodesics travelling away from the origin have $t \pm r \rightarrow \pm\infty$, i.e., $q \rightarrow \frac{1}{2}\pi$, $p \rightarrow -\frac{1}{2}\pi$ or $R \rightarrow \pi$, $T \rightarrow 0$, i.e., they approach ***spacelike infinity***, I^o, given by $R = \pi$, $T = 0$. Again the asymptotic behaviour of parallel spacelike geodesics is as stated earlier.

To study the asymptotic behaviour of null geodesics note first that conformal transformations map null vectors to null vectors. Next consider radially outgoing null geodesics $t = r + c$, i.e., $v \rightarrow \infty$, $u \rightarrow c$, where c is finite. Thus $q \rightarrow \frac{1}{2}\pi$, $p \rightarrow p_o \neq \frac{1}{2}\pi$, i.e., $T + R \rightarrow \pi$, $T - R \rightarrow 2p_o$. This is a point on the null hypersurface $T + R = \pi$ which will be called ***future null infinity*** $\mathscr{I}^+$. (The symbol $\mathscr{I}$ is pronounced "scri".) Similarly radially ingoing null geodesics originate on ***past null infinity*** $\mathscr{I}^-$ given by $T - R = -\pi$. Note that parallel geodesics have different end points.

We can of course peel the representation off the cylinder obtaining a ***Penrose diagram*** for Minkowski spacetime, fig. 3.2.2. While this is perfectly satisfactory for spherically symmetric spacetimes, in more general situations the diagram is sometimes drawn as fig. 3.2.3. However while this is perfectly satisfactory near $I^{\pm}$, it is a misleading picture of the structure near I^o, for I^o is really a point.

(3.2.3) Exercise. Locate $I^o, I^+, I^-, \mathcal{J}^+, \mathcal{J}^-$ for the compactified spacetime with $\Omega = \Omega_M$.

3.3   Asymptotic simplicity

An *isolated system* is one for which spacetime becomes more like Minkowski spacetime as one moves away from the source. Using our experience gained in section 3.2 we try to capture this idea in a definition. Unfortunately "asymptotically flat" has been defined in a number of mutually inconsistent ways in the literature, and a new name is needed to avoid confusion. (The two definitions given in this section are now universally accepted.)

(3.3.1) DEFINITION

A spacetime $(\tilde{M}, \tilde{g})$ *is* **asymptotically simple** *if there exists another manifold* (M, g) *such that*

1) $\tilde{M}$ is an open submanifold of M with smooth boundary $\partial \tilde{M}$,
2) there exists a real function Ω on M such that $g_{ab} = \Omega^2 \tilde{g}_{ab}$ on $\tilde{M}$, $\Omega = 0$, $\nabla_a \Omega \neq 0$ on $\partial \tilde{M}$,
3) every null geodesic has two end points on $\partial \tilde{M}$,
4) $\tilde{R}_{ab} = 0$ near $\partial \tilde{M}$.

$(\tilde{M}, \tilde{g})$ *is called the* **physical spacetime**, *and* (M, g) *is the* **unphysical spacetime**.

This definition is adapted from Hawking and Ellis, 1973. If condition 4 holds it is not necessary to assume that $\nabla_a \Omega \neq 0$ on $\mathcal{J}$.

The idea of conditions 1), 2) is to construct $\partial \tilde{M}$ as an "infinity" for spacetime $(\tilde{M}, \tilde{g})$, which is to have some of the properties of "infinity" for Minkowski spacetime. However it is by no means clear that we have really captured the concept of "infinity". To demonstrate this we need to consider geodesics, and so we require the connection.

(3.3.2) Exercise. If $g_{ab} = \Omega^2 \tilde{g}_{ab}$, $g^{ab} = \Omega^{-2} \tilde{g}^{ab}$, show that the Christoffel symbols are related by

$$\Gamma^a{}_{bc} = \tilde{\Gamma}^a{}_{bc} + 2\Omega^{-1}\delta^a{}_{(b}\nabla_{c)}\Omega - \Omega^{-1}(\nabla_d\Omega)\tilde{g}^{ad}\tilde{g}_{bc}. \qquad (3.3.1)$$

Let γ be a null geodesic in (M, g) with affine parameter v such that $v = 0$ on $\partial \tilde{M}$. The equation of γ is

$$\frac{d^2 x^a}{dv^2} + \Gamma^a{}_{bc}\frac{dx^b}{dv}\frac{dx^c}{dv} = 0. \qquad (3.3.2)$$

Suppose $\tilde{\gamma}$ is the corresponding geodesic in $\tilde{M}$. Using a different parameter $\tilde{v}(v)$ and equation (3.3.1), this becomes

$$\frac{d^2 x^a}{d\tilde{v}^2} + \tilde{\Gamma}^a{}_{bc}\frac{dx^b}{d\tilde{v}}\frac{dx^c}{d\tilde{v}} = -\frac{1}{\tilde{v}'}\left(\frac{\tilde{v}''}{\tilde{v}'} + 2\frac{\Omega'}{\Omega}\right)\frac{dx^a}{d\tilde{v}}, \qquad (3.3.3)$$

where a $'$ denotes a v-derivative. This is clearly the equation of $\tilde{\gamma}$, the equivalent geodesic in $\tilde{M}$. If we demand that $\tilde{v}$ be an affine parameter the right hand side must vanish. This implies $\tilde{v}' = const/\Omega^2$, and absorbing the constant into $\tilde{v}$ we obtain

$$\frac{d\tilde{v}}{dv} = \frac{1}{\Omega^2}. \qquad (3.3.4)$$

Now on $\partial \tilde{M}$, $\Omega = 0$, $\nabla_a\Omega \neq 0$, and we may choose v so that near $\partial \tilde{M}$, $v \sim -\Omega$. Thus $\tilde{v} \sim -1/v$ becomes unbounded; $\tilde{\gamma}$ never reaches $\partial \tilde{M}$, which apparently really is "at infinity" and can only be reached by conformal rescaling. This would appear to justify condition 3). Condition 4) is designed to capture the idea that spacetime shall be empty sufficiently far from the source. As we shall see in section 3.5 it can be weakened considerably.

Clearly Minkowski spacetime is asymptotically simple. Are there any other candidates? It seems reasonable that the conditions ought to include the exterior Schwarzschild spacetime

$$d\tilde{s}^2 = (1 - 2m/r)dt^2 - dr^2/(1 - 2m/r) - r^2 d\Sigma^2. \qquad (3.3.5)$$

Setting $u = t - (r + 2m \ln(r/2m - 1))$, $l = 1/r$ gives

$$d\tilde{s}^2 = l^{-2}(l^2(1 - 2ml)du^2 - 2dudl - d\Sigma^2). \qquad (3.3.6)$$

The analogous Minkowski metric $(u = t - r, \; l = 1/r)$ is

$$d\tilde{s}^2 = l^{-2}(l^2 du^2 - 2dudl - d\Sigma^2).$$

Now $\tilde{M}$ is given by $0 < l < 1/2m$. For M in the definition we may take $0 \leq l < 1/2m$, so that $\partial \tilde{M}$ is $l = 0$. We also set $\Omega = l$. Then on $\partial \tilde{M}$, $\Omega = 0$, $\nabla_a \Omega = (0, 1, 0, 0) \neq 0$. Thus conditions 1), 2), 4) are satisfied. However condition 3) cannot be satisfied, for there exist closed null geodesics at $r = 3m$ which never "escape to infinity". However as $l \to 0$ the Schwarzschild spacetime does become flat asymptotically. We therefore weaken the definition requiring it to hold essentially only in a neighbourhood of $l = 0$.

(3.3.3) DEFINITION

$(\tilde{M}, \tilde{g})$ *is weakly asymptotically simple if there exists an asymptotically simple spacetime* $(\tilde{M}', \tilde{g}')$ *and a neighbourhood* $\tilde{U}'$ *of* $\partial \tilde{M}'$ *in* $\tilde{M}'$ *such that* $\tilde{U}' \cap \tilde{M}'$ *is isometric to an open subspace* $\tilde{U}$ *of* $\tilde{M}$.

Although it may take a few seconds to understand this definition it apparently includes all known examples of spacetimes one would wish to regard as flat asymptotically.

3.4 Conformal transformation formulae

Before exploring asymptopia further it is clear that we need to establish the relationship between both the connection and curvature tensors in the physical and unphysical spacetimes. For later convenience we collect together all the relevant formulae in this section.

As suggested in section 3.1 we define

$$g_{ab} = \Omega^2 \tilde{g}_{ab}, \qquad g^{ab} = \Omega^{-2} \tilde{g}^{ab}. \qquad (3.4.1)$$

Consistent with this we choose

$$\epsilon_{AB} = \Omega \tilde{\epsilon}_{AB}, \qquad \epsilon^{AB} = \Omega^{-1} \tilde{\epsilon}^{AB}, \qquad (3.4.2)$$

plus their complex conjugates. Indeed unless we require ϵ to transform differently to $\tilde{\epsilon}$ this is the only possible choice which is continuous with the identity. (A complex Ω would lead to torsion, Penrose and Rindler, 1984.) It is important to note that we have not introduced a spin basis. Therefore equation (3.4.2), and indeed all subsequent equations, are to be understood in the abstract index formalism. When we introduce spin bases we shall take care to introduce a different notation for spinor components. We also require a covariant derivative $\nabla_{AA'}$ on the unphysical spacetime. This must satisfy the conditions of definition 2.5.1, plus equation (3.3.1). Some experimentation shows that the only plausible candidate is defined via the rules:

1) $\tilde{\nabla}_{AA'} \chi = \nabla_{AA'} \chi$ for all scalars χ,

2)

$$\tilde{\nabla}_{AA'} \xi_B = \nabla_{AA'} \xi_B + \Upsilon_{BA'} \xi_A,$$
$$\tilde{\nabla}_{AA'} \eta_{B'} = \nabla_{AA'} \eta_{B'} + \Upsilon_{AB'} \eta_{A'},$$

   where $\Upsilon_{AA'} = \nabla_{AA'} \ln \Omega$,

3)

$$\tilde{\nabla}_{AA'} \xi^B = \nabla_{AA'} \xi^B - \epsilon_A{}^B \Upsilon_{CA'} \xi^C,$$
$$\tilde{\nabla}_{AA'} \eta^{B'} = \nabla_{AA'} \eta^{B'} - \epsilon_{A'}{}^{B'} \Upsilon_{AC'} \eta^{C'},$$

   with the obvious extension to higher valence spinors,

4)

$$\tilde{\nabla}_a \xi_b = \nabla_a \xi_b + 2\xi_{(a} \Upsilon_{b)} - g_{ab} g^{cd} \xi_c \Upsilon_d.$$

The formal existence and uniqueness proof is given in Penrose and Rindler, 1984.

(3.4.1) Exercise. Show that

$$\tilde{\nabla}_{AA'} \tilde{\epsilon}_{BC} = \tilde{\nabla}_{AA'} \tilde{\epsilon}^{BC} = 0, \qquad \Longleftrightarrow \qquad \nabla_{AA'} \epsilon_{BC} = \nabla_{AA'} \epsilon^{BC} = 0.$$

(3.4.2) Exercise. Check that rule 4) is consistent both with rule 2) and with equation (3.3.1).

We need to establish a rule for index raising and lowering. If the kernel letter of a spinor carries a tilde then $\tilde{\epsilon}^{AB}$, $\tilde{\epsilon}_{AB}$ are to be used; however if there is no such tilde then the standard ϵ^{AB}, ϵ_{AB} are used.

As one might expect some key concepts turn out to be conformally invariant. As an example let $\tilde{\phi}_{AB...C}$ be a totally symmetric spinor representing a zero rest mass field. Let

$$\phi_{AB...C} = \Omega^{-1}\tilde{\phi}_{AB...C}. \tag{3.4.3}$$

Then the zero rest mass field equations are conformally invariant, i.e.

$$\tilde{\nabla}^{AA'}\tilde{\phi}_{AB...C} = 0 \qquad \Longleftrightarrow \qquad \nabla^{AA'}\phi_{AB...C} = 0. \tag{3.4.4}$$

(3.4.3) Exercise. Verify equation (3.4.4).

The transformation formulae for the various parts of the Riemann tensor are quite complicated, although considerably simpler in spinor form than their tensor counterparts. We merely state them here, relegating the derivation to a tedious exercise. They are:

$$\tilde{\Psi}_{ABCD} = \Psi_{ABCD}, \tag{3.4.5}$$

$$\tilde{\Lambda} = \Omega^2\Lambda - \tfrac{1}{4}\Omega\nabla_{CC'}\nabla^{CC'}\Omega + \tfrac{1}{2}(\nabla_{CC'}\Omega)(\nabla^{CC'}\Omega), \tag{3.4.6}$$

$$\tilde{\Phi}_{ABA'B'} = \Phi_{ABA'B'} + \Omega^{-1}\nabla_{A(A'}\nabla_{B')B}\Omega. \tag{3.4.7}$$

In vector form the last two equations are

$$\tilde{\Lambda} = \Omega^2\Lambda - \tfrac{1}{4}\Omega\Box\Omega + \tfrac{1}{2}(\nabla_c\Omega)(\nabla^c\Omega), \tag{3.4.8}$$

$$\tilde{\Phi}_{ab} = \Phi_{ab} + \Omega^{-1}\nabla_a\nabla_b\Omega - \tfrac{1}{4}\Omega^{-1}(\Box\Omega)g_{ab}, \tag{3.4.9}$$

where $\Box = \nabla_c\nabla^c$.

(3.4.4) Exercise. Obtain the relation between $\tilde{\Box}_{AB}$ and $\Box_{AB}$ acting on a spinor ξ_A or $\tilde{\xi}_{A'}$ using rule 2. Hence, using the Ricci identities (exercise 2.5.5) obtain equations (3.4.5–7) and (3.4.8–9).

3.5 $\mathscr{I}$ and peeling

Using the formulae of the last section we can now derive some of the important properties of null infinity. For convenience we set $n_a = -\nabla_a \Omega$, noting that $\nabla_{[a} n_{b]} = 0$. The boundary $\partial \tilde{M}$ of $\tilde{M}$ is given by $\Omega = 0$, and so n_a is normal to it. For the results that follow condition 4) of definition 3.3.1 for asymptotic simplicity can be relaxed in a number of ways.

(3.5.1) LEMMA

Suppose the energy momentum tensor of "physical" matter falls off sufficiently fast near $\partial \tilde{M}$

$$\tilde{T}_{ab} = O(\Omega^3) \qquad as \quad \Omega \to 0. \tag{3.5.1}$$

Then the boundary $\partial \tilde{M}$ is a null hypersurface.

Proof: Using Einstein's field equation and assumption (3.5.1), we deduce that near $\partial \tilde{M}$

$$\tilde{\Lambda} = O(\Omega), \qquad \Phi_{ab} = O(\Omega^3).$$

We see that $\tilde{\Lambda}$ vanishes on $\partial \tilde{M}$, and then it follows from equation (3.4.8) that $n_a n^a$ must vanish there. Thus $\partial \tilde{M}$ is a null hypersurface.

Locally $\tilde{M}$ must be either to the past or the future of $\partial \tilde{M}$.

(3.5.2) DEFINITION

*If $\tilde{M}$ lies to the past then the future end points of null geodesics form a null hypersurface $\mathscr{I}^+$ or **future null infinity** , while if $\tilde{M}$ lies to the future the hypersurface is **past null infinity**, $\mathscr{I}^-$.*

It can be shown, see e.g., section 6.9 of Hawking and Ellis, 1973, that each of these two hypersurfaces is topologically $S^2 \times R$.

For the next result we need a slightly stronger assumption about the matter.

(3.5.3) THEOREM

Suppose that $\tilde{T}^{ab}$ vanishes in a neighbourhood $\tilde{U}$ of $\mathscr{I}$. Then $\Psi_{ABCD} = O(\Omega)$ in $\tilde{U}$, and in particular Ψ_{ABCD} vanishes on $\mathscr{I}$.

Proof: The vacuum condition implies that $\tilde{\nabla}^{AA'}\tilde{\Psi}_{ABCD} = 0$ in $\tilde{U}$. Noting that $\Psi_{ABCD} = \tilde{\Psi}_{ABCD}$ from (3.4.5), and using (3.4.4) we have $\nabla^{AA'}(\Omega^{-1}\Psi_{ABCD}) = 0$ or

$$\Omega\nabla^{AA'}\Psi_{ABCD} = -\Psi_{ABCD}n^{AA'}, \qquad (3.5.2)$$

where $n_{AA'} = -\nabla_{AA'}\Omega$. By continuity (3.5.2) holds also at $\mathscr{I}$. But from lemma 3.5.1 n^a is null on $\mathscr{I}$ and so there exists an ι^A such that $n^a = \iota^A\bar{\iota}^{A'}$ there. It follows (c.f. theorem 2.9.4) that there exists a ψ such that

$$\Psi_{ABCD} = \psi\iota_A\iota_B\iota_C\iota_D \qquad \text{on } \mathscr{I}. \qquad (3.5.3)$$

Next we differentiate (3.5.2) again and restrict the equation to $\mathscr{I}$ obtaining

$$\iota_E\bar{\iota}_{E'}\nabla^{AA'}\Psi_{ABCD} = \iota^A\bar{\iota}^{A'}\nabla_{EE'}\Psi_{ABCD} + \Psi_{ABCD}\nabla_{EE'}\iota^A\bar{\iota}^{A'}. \qquad (3.5.4)$$

Note that (3.4.7) implies $\nabla_{E(E'}\bar{\iota}_{A')}\iota_A$ on $\mathscr{I}$. Performing this symmetrization, and using the Jacobi identity, equation (2.2.5), we find

$$\iota^E\bar{\iota}_{(E'}\nabla_{A')E}\Psi_{ABCD} = 0 \qquad \text{on } \mathscr{I},$$

which implies that

$$\iota^E\nabla_{EA'}(\psi\iota_A\iota_B\iota_C\iota_D) = 0 \qquad \text{on } \mathscr{I}. \qquad (3.5.5)$$

It will be shown later (lemma 3.9.4) that this implies that ψ vanishes on $\mathscr{I}$ and so

$$\Psi_{ABCD} = O(\Omega) \qquad \text{at } \mathscr{I}. \qquad (3.5.6)$$

This is a special case of a general result on **peeling**. Let γ be a null geodesic in M reaching $\mathscr{I}$ at the point p. Let $\tilde{\gamma}$ be the corresponding null geodesic in $\tilde{M}$. We choose a spin basis $(\tilde{o}, \tilde{\iota})$

at one point of $\tilde{\gamma}$ with $\tilde{l}$ tangent to $\tilde{\gamma}$ there, and we propagate the basis parallelly along $\tilde{\gamma}$ via

$$\tilde{D}\tilde{o}^A = \tilde{D}\tilde{\iota}^A = 0. \tag{3.5.7}$$

Here $\tilde{D} = \tilde{o}^A\tilde{\bar{o}}^{A'}\tilde{\nabla}_{AA'} = d/d\tilde{r}$, where $\tilde{r}$ is an affine parameter along $\tilde{\gamma}$. Now we are free to choose whatever conformal transformation behaviour we wish for o and ι provided that $\epsilon^{AB} = o^A\iota^B - \iota^A o^B$ transforms according to (3.4.2). The most convenient choice in this context is

$$o_A = \tilde{o}_A, \quad o^A = \Omega^{-1}\tilde{o}^A, \quad \iota_A = \Omega\tilde{\iota}_A, \quad \iota^A = \tilde{\iota}^A, \tag{3.5.8}$$

which implies

$$\begin{aligned} l_a &= \tilde{l}_a, \quad m_a = \Omega\tilde{m}_a, \quad n_a = \Omega^2\tilde{n}_a, \\ l^a &= \Omega^{-2}\tilde{l}^a, \quad m^a = \Omega^{-1}\tilde{m}^a, \quad n^a = \tilde{n}^a. \end{aligned} \tag{3.5.9}$$

Now using (3.5.7–8) and rule 3) from section 3.4 it follows that

$$Do^A = 0, \qquad D\iota^A = (\Omega^{-1}\delta\Omega)o^A. \tag{3.5.10}$$

But since n_a, l^a are normal to $\mathcal{I}$, n^a, m^a and $\bar{m}^a$ are tangent to $\mathcal{I}$, and since Ω vanishes on $\mathcal{I}$ it follows that $\Omega^{-1}\delta\Omega$ is regular there. Thus o^A, ι^A are regular on $\mathcal{I}$. Further since $Do^A = 0$, γ is affinely parametrized. We choose the origin and scaling of the parameter r to set

$$r = 0, \qquad D\Omega = \frac{d\Omega}{dr} = -1 \qquad \text{at } p. \tag{3.5.11}$$

Equation (3.5.9) implies that, acting on scalars, $D = \Omega^{-2}\tilde{D}$ or $dr = \Omega^2 d\tilde{r}$, which should be compared with equation (3.3.4). Thus from (3.5.11) we may conclude

$$\tilde{r} \sim \Omega^{-1} \qquad \text{near } \mathcal{I}. \tag{3.5.12}$$

The next result is often referred to as the **peeling theorem.**

(3.5.4) THEOREM

Suppose that $\phi_{AB\ldots C}$ represents a spin-s zero rest mass field which is regular at $\mathscr{I}$, and let

$$\phi_{(i)} = \phi_{AB\ldots C} o^A o^B \ldots \iota^C,$$

where there are i ι's and $(2s-i)$ o's. Then the physical components satisfy

$$\tilde{\phi}_{(i)} = O\left(\tilde{r}^{-(2s+1-i)}\right). \tag{3.5.13}$$

Proof: Clearly the $\phi_{(i)}$ are regular at $\mathscr{I}$. But

$$\phi_{(i)} = (\Omega^{-1}\tilde{\phi}_{AB\ldots C})(\Omega^{-1}\tilde{o}^A)(\Omega^{-1}\tilde{o}^B)\ldots(\tilde{\iota}\,{}^C_{\tilde{o}}),$$
$$= \Omega^{-(2s-i)-1}\tilde{\phi}_{(i)}.$$

Thus $\Omega^{-(2s+1-i)}\tilde{\phi}_{(i)}$ is regular at $\mathscr{I}$, in other words the physical components satisfy (3.5.13).

(3.5.5) Example. Consider electromagnetism, for which $s = 1$, $\phi_{AB\ldots C} = \phi_{AB}$. As in section 2.6 set

$$\phi_{(0)} = \phi_{AB} o^A o^B, \quad \phi_{(1)} = \phi_{AB} o^A \iota^B, \quad \phi_{(2)} = \phi_{AB} \iota^A \iota^B.$$

Then equation (3.5.13) implies that

$$\tilde{\phi}_0 \sim \tilde{r}^{-3}, \quad \tilde{\phi}_1 \sim \tilde{r}^{-2}, \quad \tilde{\phi}_2 \sim \tilde{r}^{-1},$$

or

$$\tilde{\phi}_{AB} = \frac{\tilde{\phi}_2^{(0)} \iota_A \iota_B}{\tilde{r}} + \frac{[-2\tilde{\phi}_1^{(0)} o_{(A}\iota_{B)} + \tilde{\phi}_2^{(1)}\iota_A\iota_B]}{\tilde{r}^2} + \frac{\tilde{\phi}_{AB}^{(2)}}{\tilde{r}^3} + \cdots, \tag{3.5.14}$$

where $\tilde{\phi}^{(n)}$ is $\tilde{r}$-independent. We write this schematically as

$$\tilde{\phi}_{AB} = \frac{[N]_{AB}}{\tilde{r}} + \frac{[I]_{AB}}{\tilde{r}^2} + O(\tilde{r}^{-3}), \tag{3.5.15}$$

since the leading $1/\tilde{r}$ term represents a null electromagnetic field. But as we have seen such fields correspond to plane waves. Thus far from an isolated source the leading term in the asymptotic field is a plane wave, while the next order term is algebraically general.

(3.5.6) Problem. Show that the gravitational field far from an isolated source can be written schematically as

$$\tilde{\Psi}_{ABCD} = \frac{[N]}{\tilde{r}} + \frac{[III]}{\tilde{r}^2} + \frac{[II]}{\tilde{r}^3} + \frac{[I]}{\tilde{r}^4} + O(\tilde{r}^{-5}),$$

where $[N]$ represents a Weyl tensor of type N etc.

3.6   The choice of conformal gauge

The conformal factor used to compactify spacetime has never been fixed precisely, except in examples. It can be multiplied by any real strictly positive scalar field without changing any of its properties. This "gauge freedom" can be put to good use in specific calculations. In this section we set up the key equations which can be used to determine a specific choice.

We shall make a slightly stronger assumption about the physical matter than in section 3.5, namely that in a neighbourhood of $\mathscr{I}$

$$\tilde{T}_{ab} = o(\Omega^4), \tag{3.6.1}$$

and further that this asymptotic relation can be differentiated once. It follows immediately that $\Omega\tilde{\Phi}_{ab} = o(\Omega^5)$. Then, setting $n_a = -\nabla_a\Omega$, equation (3.4.9) implies

$$\Omega\Phi_{ab} - \nabla_a n_b - \tfrac{1}{4}g_{ab}\Box\Omega = o(\Omega^5). \tag{3.6.2}$$

We now differentiate (3.6.2) obtaining

$$n_c\Phi_{ab} - \Omega\nabla_c\Phi_{ab} + \nabla_c\nabla_a n_b + \tfrac{1}{4}g_{ab}\nabla_c(\Box\Omega) = o(\Omega^4).$$

Now setting $\Omega = 0$ gives

$$n_a\Phi_{bc} + \nabla_a\nabla_b n_c + \tfrac{1}{4}g_{bc}\nabla_a(\Box\Omega) = 0 \qquad \text{on } \mathscr{I}. \tag{3.6.3}$$

Next consider identity (3.4.8) in the form

$$\Omega^2\Lambda = \tfrac{1}{4}\Omega\Box\Omega - \tfrac{1}{2}n_c n^c + o(\Omega^2). \tag{3.6.4}$$

Differentiating this equation gives

$$\Omega^2 \nabla_a \Lambda - 2\Omega \Lambda n_a = \tfrac{1}{4}\Omega \nabla_a (\Box \Omega) - \tfrac{1}{4}n_a \Box \Omega - n^c \nabla_a n_c + o(\Omega^1)$$
$$= \tfrac{1}{4}\Omega \nabla_a (\Box \Omega) - \tfrac{1}{4}n_a \Box \Omega - n^c \Omega \Phi_{ac} +$$
$$\tfrac{1}{4}n_a \Box \Omega + o(\Omega^1),$$

where equation (3.6.2) has been used. We now divide this equation by Ω and subsequently set $\Omega = 0$ obtaining

$$2n_a \Lambda = -\tfrac{1}{4}\nabla_a (\Box \Omega) + n^b \Phi_{ab} \qquad \text{on } \mathscr{I}. \qquad (3.6.5)$$

We can now explore the freedom left in Ω. As stated above we should consider the changes

$$\Omega \mapsto \widehat{\Omega} = \theta \Omega, \qquad\qquad\qquad (3.6.6)$$

where $\theta \neq 0$ on $\mathscr{I}$. (This condition arises because we have postulated $\nabla_a \Omega \neq 0$ on $\mathscr{I}$. If θ vanished there then Ω would vanish to second order.)

(3.6.1) LEMMA

Under a conformal gauge change (3.6.6) we have on $\mathscr{I}$

$$\widehat{n}_b = \theta n_b, \qquad \widehat{\nabla}_a \widehat{n}_b = \theta \nabla_a n_b + g_{ab} n^c \nabla_c \theta. \qquad (3.6.7)$$

Proof: Set $g_{ab} = \Omega^2 \tilde{g}_{ab}$ and $\widehat{g}_{ab} = \widehat{\Omega}^2 \tilde{g}_{ab}$ so that $g_{ab} = \theta^{-2}\widehat{g}_{ab}$. Then setting $t_b = \nabla_b \theta$, we find

$$\widehat{n}_b = -\widehat{\nabla}_b \widehat{\Omega} = -\nabla_b \widehat{\Omega}$$
$$= -\nabla_b (\theta \Omega) = \theta n_b - \Omega t_b,$$

where rule 1 of section 3.4 has been used. Rule 4 implies

$$\widehat{\nabla}_a \widehat{n}_b = \nabla_a \widehat{n}_b - 2t_{(a}\widehat{n}_{b)}/\theta + g_{ab}g^{cd}\widehat{n}_c t_d/\theta$$
$$= \theta \nabla_a n_b + t_a n_b + n_a t_b - \Omega \nabla_a t_b - 2t_{(a}n_{b)} +$$
$$2(\Omega/\theta)t_a t_b + g_{ab}g^{cd}[n_c t_d - (\Omega/\theta)t_c t_d].$$

These equations simplify considerably on $\mathscr{I}$ giving (3.6.7).

Lemma 3.6.1 is the key to the choice of conformal gauge. Perhaps the most useful choice for θ is that which sets $\widehat{\nabla}_a \widehat{n}_b = 0$. To see how this is possible note that equation (3.6.2) implies $\frac{1}{4} g_{ab} \Box \Omega = -\nabla_a n_b$ on $\mathscr{I}$. Hence on $\mathscr{I}$

$$\begin{aligned}
\tfrac{1}{4} \widehat{g}_{ab} \widehat{\Box} \widehat{\Omega} &= -\widehat{\nabla}_a \widehat{n}_b \\
&= -\theta \nabla_a n_b - g_{ab} n^c \nabla_c \theta \\
&= \tfrac{1}{4} g_{ab} (\theta \Box \Omega - 4\Delta\theta).
\end{aligned}$$

Now Δ is the directional derivative along the generators of $\mathscr{I}$. Suppose that on some initial 2-surface S_o in $\mathscr{I}$ transverse to the generators we choose θ arbitrarily. We can then propagate θ via

$$\Delta\theta = (\tfrac{1}{4}\Box\Omega)\theta$$

on $\mathscr{I}$. This sets $\widehat{\nabla}_a \widehat{n}_b = 0$ on $\mathscr{I}$. We shall assume that this has been done, and henceforth drop the carets. The freedom in the choice of θ on S_o can be expressed as residual gauge freedom

$$\Omega \mapsto \widehat{\Omega} = \theta\Omega, \qquad \Delta\theta = 0 \qquad \text{on } \mathscr{I}, \tag{3.6.8}$$

and will be used later.

3.7   A spin basis adapted to $\mathscr{I}$

Let us recall that we have defined $n_a = -\nabla_a \Omega$ where n^a is on $\mathscr{I}$, and have used the conformal gauge freedom to set

$$\nabla_a n_b = 0 \qquad \text{on } \mathscr{I}.$$

Clearly there exists a spinor ι defined in a neighbourhood of $\mathscr{I}$ such that

$$n_a = \iota_A \bar{\iota}_{A'}, \qquad \nabla_a(\iota_B \bar{\iota}_{B'}) = 0 \qquad \text{on } \mathscr{I}.$$

We can complete ι to a spin basis (o, ι) in this neighbourhood. Now on $\mathscr{I}$

$$0 = \bar{m}^b \nabla_a n_b = \iota^B \nabla_a \iota_B,$$

and so there exists a β_a such that

$$\nabla_a \iota_B = i\beta_a \iota_B, \tag{3.7.1}$$

there. Further $\nabla_a n_b = 0$ implies $i(\beta_a - \bar\beta_a) = 0$ so that β_a must be real. Based on our earlier experience in the proof of theorem 2.9.4 it seems worthwhile to explore the possibility of making a flag rotation or phase change $\iota \mapsto \hat\iota = e^{-i\phi}\iota$ so as to set

$$\nabla_a \hat\iota_B = 0. \tag{3.7.2}$$

As we found earlier, it may take some time to prove an apparently subsidiary result but this is ultimately time saved!

(3.7.1) THEOREM

A phase change can always be found to ensure that $\hat\iota$ is covariantly constant on $\mathscr{I}$.

Proof:
   It is clear that (3.7.2) will hold on $\mathscr{I}$ iff

$$\nabla_a \hat\iota_B = i(\beta_a - \nabla_a \phi)\hat\iota_B,$$

there. Thus in order to obtain (3.7.2) on $\mathscr{I}$ it is necessary and sufficient to be able to solve

$$\nabla_a \phi = \beta_a, \tag{3.7.3}$$

there. On $\mathscr{I}$ β must be a gradient, but we cannot conclude that this has to hold in a neighbourhood of $\mathscr{I}$, since Ω vanishes at $\mathscr{I}$. The integrability condition for (3.7.3) has to be weakened to

$$\beta_a = \nabla_a \lambda + \Omega v_a, \tag{3.7.4}$$

in a neighbourhood of $\mathscr{I}$ for some vector field v, and some scalar field λ. An equivalent formulation of the condition (3.7.4) is that on $\mathscr{I}$

$$\nabla_{[a}\beta_{b]} = n_{[a}v_{b]}. \tag{3.7.5}$$

There is an equivalent spinor formulation. Equation (3.7.1) can hold on $\mathscr{I}$ if and only if there exists a spinor field γ_{bC} defined in a neighbourhood of $\mathscr{I}$ such that

$$\nabla_b \iota_C = i\beta_b \iota_C + \Omega\gamma_{bC},$$

there. This equation implies

$$\nabla_a \nabla_b \iota_C = i(\nabla_a \beta_b + \beta_a \beta_b)\iota_C - n_a \gamma_{bC},$$

on $\mathscr{I}$. Thus an equivalent formulation of the integrability condition (3.7.5) for equation (3.7.3) is that on $\mathscr{I}$

$$\nabla_{[a} \nabla_{b]} \iota_C = n_{[a} L_{b]C}, \tag{3.7.6}$$

for some spinor field $L_{BCB'}$

We now recall that on $\mathscr{I}$ Ψ_{ABCD} vanishes, and so, from equation (2.5.6) and excerise 2.5.5

$$2\nabla_{[a} \nabla_{b]} \iota_C = \epsilon_{A'B'} \square_{AB} \iota_C + \epsilon_{AB} \square_{A'B'} \iota_C$$
$$= -2\Lambda\iota_{(A} \epsilon_{B)C} \epsilon_{A'B'} + \epsilon_{AB} \Phi_{CDA'B'} \iota^D.$$

Thus the integrability condition (3.7.6) on $\mathscr{I}$ can be reformulated as

$$-2\Lambda\iota_{(A} \epsilon_{B)C} \epsilon_{A'B'} + \epsilon_{AB} \Phi_{CDA'B'} \iota^D = 2n_{[a} L_{b]C}, \tag{3.7.7}$$

there.

Although it may seem somewhat inelegant it is advantageous to extract the 32 scalar equations in (3.7.7). It is clear that on multiplication by $\iota^A \iota^B$ the equation becomes trivial, while multiplying it by $o^A \iota^B$ is equivalent to multiplying it by $\iota^A o^B$. The 16 remaining equations either define consistently components of $L_{BCB'}$ or produce the equations

$$\Phi_{12} = \Phi_{22} = \Phi_{11} + \Lambda = 0.$$

The vanishing of Φ_{12}, Φ_{22} is already guaranteed by equation (3.6.5) since $\square\,\Omega = 0$ on $\mathscr{I}$. Thus the integrability condition (3.7.7) reduces to the single scalar equation

$$\Phi_{11} + \Lambda = 0. \tag{3.7.8}$$

Can we satisfy it?

We are assuming $\mathscr{I}$ is topologically $S^2 \times R$. Let S be a *cut of* $\mathscr{I}$, i.e., a spacelike 2-surface in $\mathscr{I}$ orthogonal to the generators of $\mathscr{I}$. There exists another null hypersurface Σ_S (at least locally) which contains S and whose generators are orthogonal to S. We use Σ_S to define a spinor field o on S as follows:

i) $l^a = o^A \bar{o}^{A'}$ is tangent to the generators of Σ_S,
ii) $o_A \iota^A = 1$.

It follows immediately that $\kappa = 0$ on S. We may propagate o from S onto $\mathscr{I}$ via the condition $\Delta o^A = 0$, which implies that $\tau = \gamma = 0$ on $\mathscr{I}$. We have yet to use the restriction $\nabla_a n_b = 0$ on $\mathscr{I}$ which produces

$$\lambda = \mu = \nu = \pi = \bar{\alpha} + \beta = \epsilon + \bar{\epsilon} = 0. \qquad (3.7.9)$$

on $\mathscr{I}$. Field equation (q) of appendix B implies

$$\Delta \rho = -2\Lambda. \qquad (3.7.10)$$

Now the assertion that ρ is real is equivalent to the assertion that l is hypersurface orthogonal, cf. equation (2.7.6). By construction l is hypersurface orthogonal on S, so that ρ is real there. But then equation (3.7.10) implies that ρ is real on $\mathscr{I}$, and so l is hypersurface orthogonal at each point of $\mathscr{I}$. Thus our initial cut/hypersurface, S/Σ_S, defines a family of cuts/hypersurfaces on $\mathscr{I}$. Now from the D, Δ commutator we deduce immediately that

$$[n, l] = (\gamma + \bar{\gamma})l + (\epsilon + \bar{\epsilon})n - (\bar{\tau} + \pi)m - (\tau + \bar{\pi})\overline{m}.$$

which vanishes on $\mathscr{I}$. Thus this family of cuts/hypersurfaces can be obtained by Lie-dragging S/Σ_S along $\mathscr{I}$. On S, the 2-metric is $g_{ab}^{(2)} = -2m_{(a}\overline{m}_{b)}$ and the 2-connection is essentially $m^a \delta m_a = \beta - \bar{\alpha}$. The 2-curvature tensor can have only one real independent component, and the defining equation for $R^{(2)}$ must be of the form $\delta \alpha - \bar{\delta}\beta +$ 4 terms quadratic in the connection. Inspection of field equation (1) of appendix B shows that to within a factor $R^{(2)}$ must be $\Phi_{11} + \Lambda$. Thus the integrability condition (3.7.8) is that the cut S, and those obtained by Lie-dragging it, be flat. Can this be satisfied?

Now the remaining conformal freedom is given by equation (3.6.8) viz.

$$\Omega \to \theta\Omega, \qquad \Delta\theta = 0 \qquad \text{on } 𝒥,$$

with θ arbitrary on S. Under such a transformation

$$R^{(2)} \to \theta^2 R^{(2)} + 2\theta\Box\theta - 2\nabla^a\theta\nabla_a\theta.$$

(This is essentially equation (3.4.8) but with different, dimension-dependent coefficients.) We can always choose θ on S to ensure that $R^{(2)}$ vanishes there. Now

$$g_{ab}^{(2)} = g_{ab} - 2l_{(a}n_{b)}.$$

Also $\nabla_{(a}n_{b)} = 0$ implies $\mathcal{L}_n g_{ab} = 0$, and by construction $\mathcal{L}_n l^a = 0$. Thus $\mathcal{L}_n g_{ab}^{(2)} = 0$ on $𝒥$ and so $\mathcal{L}_n R^{(2)} = 0$ there. Thus our choice of θ on S has ensured that each of the cuts of $𝒥$ is flat, i.e., the integrability condition (3.7.10) for equation (3.7.3) can be satisfied.

Thus we may always choose

$$\nabla_a l_A = 0 \qquad \text{on } 𝒥. \qquad (3.7.11)$$

This ensures that in addition to (3.7.9) above we have

$$\epsilon = \alpha = \beta = 0 \qquad \text{on } 𝒥. \qquad (3.7.12)$$

However there is still some conformal gauge freedom left! We may set

$$\Omega \to \widehat{\Omega} = F\Omega, \qquad \text{with } F = 1 \quad \text{on } 𝒥. \qquad (3.7.13)$$

Under (3.7.13) n^a and its gradient change, but as examination of equation (3.6.7) and the argument leading to it shows, the condition $F = 1$ on $𝒥$ implies

$$\widehat{n}_a = n_a, \qquad \widehat{\nabla}_a\widehat{n}_b = \nabla_a n_b \qquad \text{on } 𝒥.$$

Thus nothing that we have done already on $𝒥$ is affected by a conformal gauge change (3.7.13). Next we set $\widehat{l}_a = l_a$ consistent with equation (3.5.9). Since l is affinely parametrized $\rho = -\frac{1}{2} g^{ab} \nabla_a l_b$, and

$$\hat{\rho} = \frac{(\rho - F^{-1} DF)}{F^2}.$$

Clearly F can be chosen to set $\hat{\rho} = 0$ and this in addition forces $D\hat{\rho} = 0 = \hat{D}\hat{\rho}$.

Thus all of the NP connection scalars except σ vanish on $\mathscr{I}$. In fact σ contains all of the dynamical information regarding the outgoing radiation field. (The gravitational peeling theorem shows that on $\mathscr{I}$ the dominant $1/r$ term in $\tilde{\Psi}_{ABCD}$ is $\tilde{\Psi}_4$ and we regard this as determining the **outgoing field**. If the roles of l, n were interchanged the dominant term would be $\tilde{\Psi}_0$ and so we regard Ψ_0 as describing the **incoming** field.) To see the rôle of σ note that since Ψ_{ABCD} vanishes on $\mathscr{I}$ the field equations (a,k,p,q) of appendix B imply

$$\Phi_{00} = -\sigma\bar{\sigma}, \quad \Phi_{01} = -\bar{\delta}\sigma, \quad \Phi_{02} = -\Delta\sigma, \quad \Phi_{11} = \Lambda = 0,$$

and in addition equation (b) implies that $D\sigma = 0$. In older accounts $\Phi_{20} = -\Delta\bar{\sigma}$ is called the **news function, N**.

We can extract some further information intrinsic to $\mathscr{I}$ from the Bianchi identities. We first set

$$\Psi_{ABCD} = \Omega \chi_{ABCD},$$

and note that since from equation (3.5.11) we may take $D\Omega = -1$ on $\mathscr{I}$, then $D\Psi_n = -\chi_n$ there. The Bianchi identity (Bg) of appendix B implies immediately

$$\chi_4 = -\Delta^2 \bar{\sigma}. \tag{3.7.14}$$

Equations (Bj,k) imply that $D\Phi_{12} = D\Phi_{22} = 0$ so that these quantities vanish to second order. Now equation (Be) implies that

$$\chi_3 = -\delta\Delta\bar{\sigma}. \tag{3.7.15}$$

Finally although $D\Lambda$ need not vanish on $\mathscr{I}$, it must be real, and so (Bc) implies

$$\chi_2 + \sigma\Delta\bar{\sigma} + \delta^2\bar{\sigma} = \bar{\chi}_2 + \bar{\sigma}\Delta\sigma + \bar{\delta}^2\sigma. \tag{3.7.16}$$

We are therefore justified in regarding σ as measuring the outgoing asymptotic gravitational field. The other NP quantities are "pure gauge".

3.8   Asymptotic symmetry

In the previous section the cuts of $\mathscr{I}$ were locally flat 2-surfaces, i.e., locally coordinates x, y could be found so that the line element of the cut was

$$ds^2 = -dx^2 - dy^2.$$

We must emphasise however that this is deceptive globally, for each cut is topologically a S^2. We may however include the global topology at minimal cost by regarding the cut as a Riemann sphere with independent coordinates $(\zeta, \bar{\zeta})$, where

$$\zeta = x + iy, \qquad \bar{\zeta} = x - iy,$$

and

$$ds^2 = -d\zeta d\bar{\zeta}. \tag{3.8.1}$$

Although this may appear merely a cosmetic change, it is rather more than this. For we may now make an (inverse) *stereographic projection*

$$\zeta = e^{i\phi} \cot \tfrac{1}{2}\theta,$$

finding after some algebra

$$ds^2 = -\tfrac{1}{4}(1 + \zeta\bar{\zeta})^2 d\Sigma^2, \qquad d\Sigma^2 = d\theta^2 + \sin^2\theta d\phi^2. \tag{3.8.2}$$

Thus by a conformal transformation with factor P^{-1} where $P = \tfrac{1}{2}(1 + \zeta\bar{\zeta}) = \tfrac{1}{2}\mathrm{cosec}\,\tfrac{1}{2}\theta$ each cut becomes the unit sphere. We shall use this alternative representation frequently.

The construction of section 3.7 established that given one cut there was a preferred spin frame and system of cuts adapted to that cut. We shall call this a *Bondi system* based on that cut. The only thing that we have not fixed is the coordinate chart. Covariance implies that all charts are equally good. However we may prefer those adapted to the above structure, and this can be used to define an asymptotic symmetry group.

Consider first a relabelling of the generators $\zeta \mapsto \widehat{\zeta} = \widehat{\zeta}(\zeta, \bar{\zeta})$. If we require the transformation to be both conformal and locally proper, then we must require

$$d\widehat{\zeta} = C(\zeta, \bar{\zeta})d\zeta.$$

Writing $\widehat{\zeta} = \hat{x} + i\hat{y}$ we find immediately that

$$\frac{\partial \hat{x}}{\partial x} = \frac{\partial \hat{y}}{\partial y}, \qquad \frac{\partial \hat{x}}{\partial y} = -\frac{\partial \hat{y}}{\partial x},$$

so that $\widehat{\zeta} = f(\zeta)$ where f is holomorphic. This fixes the form of f, and since the result may not be known we present it formally as a lemma.

(3.8.1) LEMMA

Let f be a holomorphic bijection from the Riemann sphere to itself. Then f is a fractional linear transformation (FLT)

$$\widehat{\zeta} = f(\zeta) = \frac{a\zeta + b}{c\zeta + d}, \tag{3.8.3}$$

where, without loss of generality, $ad - bc = 1$. (In fact provided f is not constantly infinite we may drop the bijectivity requirement.)

Proof: Let T be a FLT which maps $f(0)$ to 0 and $f(\infty)$ to ∞. Then $g = Tf$ is a bijection which leaves 0 and ∞ invariant. If $g'(0) = 0$ then g could not be a bijection in a neighbourhood of 0 and so we may conclude that $g(\zeta) = \zeta h(\zeta)$ with $h(0)$ non-zero. However

$$\tilde{g}(\zeta) = g(\zeta^{-1})^{-1} = \frac{\zeta}{h(\zeta^{-1})},$$

is also a holomorphic bijection mapping 0 to 0. Thus

$$\frac{\tilde{g}(\zeta)}{\zeta} = \frac{1}{h(\zeta^{-1})},$$

has a non-zero limit as $\zeta \to 0$. Thus h is a bounded function and by Liouville's theorem h is constant. It follows that $Tf(\zeta) = k\zeta$ for some constant k and so f must be a FLT. The normalization is conventional.

It follows that if we wish a cut to remain the unit sphere under (3.8.3) then we must make another conformal transformation

$$d\widehat{\Sigma}^2 = K^2 d\Sigma^2,$$

where $K(\zeta, \bar{\zeta})$ can be determined from the equations above as

$$K = \frac{1 + \zeta\bar{\zeta}}{(a\zeta + b)(a\bar{\zeta} + \bar{b}) + (c\zeta + d)(c\bar{\zeta} + \bar{d})}. \qquad (3.8.4)$$

Thus lengths within the cut scale by a factor K under the transformation (3.8.3). For our theory to remain invariant under this rescaling "lengths" along the generators of f must scale by the same amount, i.e., we need to impose

$$d\widehat{u} = K du,$$

which integrates to give

$$\widehat{u} = K\left[u + \alpha(\zeta, \bar{\zeta})\right]. \qquad (3.8.5)$$

The transformations (3.8.3,5) together with formula (3.8.4) clearly form a group, the *Bondi-Metzner-Sachs group (BMS)* or *asymptotic symmetry group*. The largest proper normal subgroup is the *supertranslation group (S)*

$$\widehat{u} = u + \alpha(\zeta, \bar{\zeta}), \qquad \widehat{\zeta} = \zeta. \qquad (3.8.6)$$

(The reason for the name will appear below.) The factor group BMS/S is the group of conformal transformations of the sphere (3.8.3), which is isomorphic to $SL(2, \mathcal{C})$, and represents the proper orthochronous Lorentz group. There is also a 4-parameter normal subgroup, the *translation group (T)* given by (3.8.6) with

$$\alpha = \frac{A + B\zeta + \bar{B}\bar{\zeta} + C\zeta\bar{\zeta}}{1 + \zeta\bar{\zeta}}, \qquad (3.8.7)$$

where A, C are real.

(3.8.2) Problem. Let t, x, y, z be Cartesian coordinates in Minkowski space-time and set $u = t - r, r^2 = x^2 + y^2 + z^2$. Introducing $\zeta = e^{i\phi} \cot \frac{1}{2}\theta$ show that

$$Z^2\zeta = (x + iy)(1 - z/r)/4r, \quad x = r(\zeta + \bar{\zeta})Z,$$

$$y = -ir(\zeta - \bar{\zeta})Z, \quad z = r(\zeta\bar{\zeta} - 1)Z,$$

where $Z = 1/(1 + \zeta\bar{\zeta})$. Now make a translation

$$t \mapsto t' = t + a, \quad x \mapsto x' = x + b, \quad y \mapsto y' = y + c, \quad z \mapsto z' = z + d.$$

Show that

$$u \mapsto u' = u + Z(A + B\zeta + \bar{B}\bar{\zeta} + C\zeta\bar{\zeta}) + O(1/r),$$

$$\zeta \mapsto \zeta' = \zeta + O(1/r),$$

with $A = a + d$, $B = b - ic$ and $C = a - d$. Thus a real translation in Minkowski space-time generates a member of $\mathcal{T}$, thereby justifying the name. The "supertranslation group" contains all of these, and of course very many more.

(3.8.3) Problem. Working in a frame in which the Bondi cuts are flat, show that under a supertranslation (3.8.6)

$$\hat{\sigma}(\hat{u}, \zeta, \bar{\zeta}) = \sigma(\hat{u} - \alpha(\zeta, \bar{\zeta}), \zeta, \bar{\zeta}) + \delta^2\alpha(\zeta, \bar{\zeta}).$$

A *good cut* is one on which $\sigma = 0$, e.g., that produced by the null cone of a point in Minkowski spacetime. (A Bondi system of good cuts can be found in any stationary asymptotically simple spacetime.) Show using problem 3.8.2 plus a conformal transformation, that in Minkowski spacetime the subgroup of supertranslations preserving good cuts is precisely $\mathcal{T}$. Next consider the following model. An isolated system is initially stationary, it starts to radiate and then stops, so that the spacetime eventually becomes stationary again. Are the initial and final Bondi systems of good cuts related by a supertranslation, and if so is it a translation?

A symmetry group of considerable significance in special relativity is the Poincaré group. A Poincaré transformation may be regarded as the composition of a Lorentz transformation and a translation. Similarly a BMS transformation may be considered as the composition of a Lorentz transformation and a supertranslation. Thus there are many Poincaré groups at $\mathscr{I}$, indeed one for each supertranslation which is not a translation, and no one of them is preferred. This causes considerable difficulties, e.g., there is no generally accepted definition of angular momentum.

3.9 Spin-weighted spherical harmonics

In order to simplify the subsequent discussion we shall introduce the machinery of spin-weighted spherical harmonics. These have plenty of other uses as well, particularly in the representation theory of $O(4)$. The standard reference is Goldberg et al., 1967, but in order to conform to the conventions of Penrose and Rindler, 1984, some changes in the notation have been made. These will be described later.

Let S be a sphere of radius R with line element

$$ds^2 = -R^2(d\theta^2 + \sin^2\theta \, d\phi^2) = -2m_{(a}\overline{m}_{b)}dx^a dx^b, \qquad (3.9.1)$$

where $m, \overline{m}$ span $T(S)$, are null and $m_a\overline{m}^a = -1$. Note that this does not fix $m, \overline{m}$ for there is the "gauge" freedom

$$m \mapsto \widehat{m} = e^{i\psi}m, \qquad \overline{m} \mapsto \widehat{\overline{m}} = e^{-i\psi}\overline{m}, \qquad (3.9.2)$$

where ψ is real but otherwise arbitrary. This is called a **spin through** ψ. We may project space-time vectors and tensors into S using the projection tensor P: the projection of u^a is $P_b{}^a u^b$ where $P_b{}^a = -m^a\overline{m}_b - \overline{m}^a m_b$. (Projection simply discards components transverse to S.) Let $\eta_{a...by...z}$ be a tensor in S, i.e., it has been projected into S on every index, where there are p indices in the first index set and q in the second. Let

$$\eta = m^a \ldots m^b \overline{m}^y \ldots \overline{m}^z \eta_{a...by...z}.$$

It is clear that under spins (3.9.2) the scalar η transforms as

$$\eta \mapsto \widehat{\eta} = \eta e^{is\psi}, \qquad (3.9.3)$$

where $s = p - q$ is the **spin weight** of η.

By projecting the vector equivalents of equations (2.6.11) (which are given explicitly in appendix B) into S we obtain

$$\delta m^a = -(\bar{\alpha} - \beta)m^a, \quad \delta \overline{m}^a = (\bar{\alpha} - \beta)\overline{m}^a, \quad \text{in } S.$$

We can now define new derivations in S, viz $\eth$, $\overline{\eth}$, pronounced "edth, edth bar" by

$$\begin{aligned}
\eth\eta &= m^a \dots m^b \overline{m}^y \dots \overline{m}^z m^c \nabla_c \eta_{a\dots by\dots z}, \\
\overline{\eth}\eta &= m^a \dots m^b \overline{m}^y \dots \overline{m}^z \overline{m}^c \nabla_c \eta_{a\dots by\dots z},
\end{aligned} \qquad (3.9.4)$$

where ∇_c is the covariant derivative in S, or equivalently the projection of the space-time covariant derivative. It is easy to see that if η has spin weight s then $\eth\eta$ has spin weight $s + 1$, and $\overline{\eth}\eta$ has spin weight $s - 1$. Thus acting on quantities with a definite spin weight $\eth, \overline{\eth}$ can be allocated spin weights of 1 and -1 respectively.

(3.9.1) Exercise. Show that acting on a quantity η of spin weight s, $\eth$ and δ are related by

$$\eth\eta = \delta\eta + s(\bar{\alpha} - \beta)\eta, \quad \overline{\eth}\eta = \delta\eta - s(\alpha - \bar{\beta})\eta. \qquad (3.9.5)$$

Thus neither δ nor α, β have spin weight, i.e., they do not transform homogeneously under a spin.

$\eth$ and $\overline{\eth}$ can be expressed explicitly in terms of polar coordinates (θ, ϕ) or complex coordinates $\zeta, \bar{\zeta}$. It is easy to obtain the Christoffel symbols for the line element (3.9.1) as

$$\Gamma^\theta_{\phi\phi} = -\sin\theta\cos\theta, \quad \Gamma^\phi_{\theta\phi} = -\cot\theta.$$

An appropriate choice for $m, \overline{m}$ is

$$m^a = 2^{-1/2}R^{-1}(1, -i\,\text{cosec}\,\theta), \quad m_a = -2^{-1/2}R(1, -i\sin\theta), \qquad (3.9.6)$$

which is consistent with (3.9.1). Then a trivial calculation gives

$$\alpha = -\beta = -\frac{\cot\theta}{2^{3/2}R}.$$

Now we can write (3.9.5) **explicitly as**

$$\eth\eta = \frac{(\sin\theta)^s}{\sqrt{2}R}\left(\frac{\partial}{\partial\theta} - \frac{i}{\sin\theta}\frac{\partial}{\partial\phi}\right)\left[(\sin\theta)^{-s}\eta\right],$$

$$\bar{\eth}\eta = \frac{(\sin\theta)^{-s}}{\sqrt{2}R}\left(\frac{\partial}{\partial\theta} + \frac{i}{\sin\theta}\frac{\partial}{\partial\phi}\right)\left[(\sin\theta)^s\eta\right]. \qquad (3.9.7)$$

The operator $\eth\bar{\eth}$ clearly has spin weight zero, but it is not the same as $\bar{\eth}\eth$. It is easy to derive the commutator relation.

(3.9.2) Exercise. Show that if η has spin weight s then

$$(\bar{\eth}\eth - \eth\bar{\eth})\eta = \frac{s}{R^2}\eta. \qquad (3.9.8)$$

With this representation it is possible to define an orthogonal set of functions which are complete on the sphere, and are indeed eigenfunctions of $\eth\bar{\eth}$. The conventional spherical harmonics will be denoted $Y_{lm}(\theta,\phi)$. The **spin-weighted spherical harmonics** are defined as follows

$$_0Y_{lm}(\theta,\phi) = Y_{lm}(\theta,\phi), \qquad (3.9.9)$$

$$_sY_{lm}(\theta,\phi) = \begin{cases} (-1)^s R^s \left[2^s\frac{(l-s)!}{(l+s)!}\right]^{1/2} \eth^s Y_{lm}(\theta,\phi) & 0 \le s \le l, \\[2ex] R^{-s}\left[2^{-s}\frac{(l+s)!}{(l-s)!}\right]^{1/2} \bar{\eth}^{-s}Y_{lm}(\theta,\phi) & 0 \ge s \ge -l, \\[2ex] 0 & \text{otherwise.} \end{cases}$$
$$(3.9.10)$$

(3.9.3) Exercise. Show explicitly that $\eth, \bar{\eth}$ act as spin raising and lowering operators. Verify that $_sY_{lm}(\theta,\phi)$ is an eigenfunction of $\bar{\eth}\eth$. What is the corresponding eigenvalue?

Although for a given spin weight s the $_sY_{lm}$ are complete, some care is needed in changing spin weights, as can be seen in the following lemma.

(3.9.4) LEMMA

Suppose η is continuous on the sphere and has spin weight $s > 0$. Then if $\bar{\eth}\eta = 0$ it follows that $\eta = 0$. However $\eth\eta = 0$ implies *only that η is a linear combination of the $_sY_{sm}$. The equivalent results for $s < 0$ are obtained by interchanging $\eth$ and $\bar{\eth}$ in the above assertions.*

Proof: By completeness we may express η as

$$\eta = \sum_{l=s}^{\infty} \sum_{m=-l}^{l} c_{slm}\,{}_sY_{lm}. \tag{3.9.11}$$

Differentiating (3.9.11) gives

$$\bar{\eth}\eta = \sum_{l=s}^{\infty} \sum_{m=-l}^{l} a_{slm}c_{slm}\,{}_{s-1}Y_{lm}, \tag{3.9.12}$$

$$\eth\eta = \sum_{l=s}^{\infty} \sum_{m=-l}^{l} b_{slm}c_{slm}\,{}_{s+1}Y_{lm}, \tag{3.9.13}$$

where the non-vanishing factors a_{slm}, b_{slm} can be deduced from (3.9.10). Next suppose that $\bar{\eth}\eta = 0$. By completeness each term in the sum (3.9.12) must vanish. Since neither a_{slm} nor $_{s-1}Y_{lm}$ vanish for $l \geq s$, it follows that each c_{slm} vanishes, i.e., $\eta = 0$. However a similar argument applied to (3.9.13) has to be modified; since $_{s+1}Y_{sm}$ vanishes identically the c_{ssm} are undetermined. A similar argument can be used if $s < 0$.

As an immediate application of this lemma we complete the argument, started in section 3.5, that the Weyl spinor vanishes on $\mathscr{I}$. It is clear that the field ψ defined by (3.5.3) has spin weight $s = -2$, since Ψ_{ABCD} cannot carry spin weight. Also transvecting equation (3.5.5) with $o^{A'}$ gives $\bar{\eth}\psi = 0$ on $\mathscr{I}$. Thus by the above lemma $\psi = 0$ on $\mathscr{I}$ which implies (3.5.6).

We can also write $\eth, \bar{\eth}$ in terms of complex coordinates such as $\zeta, \bar{\zeta}$. A complex coordinate ξ is said to be **holomorphic** if the lines $\Re(\xi) = const.$ are orthogonal to the lines $\Im(\xi) = const.$, the spacing of each family is the same, and a rotation through $\frac{1}{2}\pi$ converts the $\Re(\xi)$-increasing direction to the $\Im(\xi)$- increasing one. Clearly a holomorphic function of a holomorphic coordinate is also a holomorphic coordinate. An obvious example on the Riemann sphere is $\xi = x + iy$. Now from section 2.3

$$\overline{m}_a = 2^{-\frac{1}{2}}(e_{1_a} + ie_{2_a}).$$

Thus if ξ is holomorphic we expect $\nabla_a \xi$ to be parallel to $\overline{m}_a$. Noting that $m^a m_a = 0$, $\overline{m}^a m_a = -1$ and $d\xi(\partial/\partial\xi) = 1$, $d\overline{\xi}(\partial/\partial\xi) = 0$, we may write

$$m^a \nabla_a = P\frac{\partial}{\partial\xi}, \qquad \overline{m}_a dx^a = -\frac{1}{P}d\xi, \qquad (3.9.14)$$

where we shall assume that $P(\xi, \overline{\xi})$ is real. (This restriction is easily removed.) Then

$$ds^2 = -2m_{(a}\overline{m}_{b)}dx^a dx^b = -\frac{2d\xi d\overline{\xi}}{P^2}. \qquad (3.9.15)$$

One might have expected $\zeta = e^{i\phi}\cot\frac{1}{2}\theta$ to be a holomorphic coordinate on the unit sphere. In that case $\xi = \ln\zeta = \ln(\cot\frac{1}{2}\theta) + i\phi$ would also be holomorphic. However

$$d\xi = -\operatorname{cosec}\theta d\theta + i\,d\phi.$$

Thus the lines $\Re(\xi) = const.$ are lines of constant latitude with $\Re(\xi)$ increasing towards the north pole, while $\Im(\xi)$ measures longitude. This would be holomorphic if we took the normal to the sphere to point inwards. Current conventions require the normal to point outwards, and so a suitable candidate for a holomorphic coordinate is

$$\xi = -\ln\overline{\zeta} = \ln(\tan\frac{1}{2}\theta) + i\phi, \quad d\xi = \operatorname{cosec}\theta\,d\theta + i\,d\phi. \quad (3.9.16)$$

It is however more convenient to take $\overline{\zeta}$ as our holomorphic coordinate ξ. Comparing equations (3.8.1,2) and (3.9.15) now shows that for a sphere of radius R and outward normal

$$P(\zeta, \overline{\zeta}) = \frac{(1 + \zeta\overline{\zeta})}{\sqrt{2}R}. \qquad (3.9.17)$$

If we had chosen $\xi = -\ln\zeta$ as our holomorphic coordinate then
the m^a defined by (3.9.14) would be parallel to the m^a defined by
(3.9.6), and the formula corresponding to (3.9.7) could have been
read off directly. Instead we proceed as follows. Equation (3.9.5)
implies that acting on a scalar of vanishing spin weight, $\eth = \delta$.
Then (3.9.14) implies that $P = \partial\bar\zeta$ since ζ has no spin weight. We
shall require the commutator relation (3.9.8) to hold and so

$$\bar\eth P = \bar\eth\eth\bar\zeta = \eth\bar\eth\bar\zeta = 0.$$

Now suppose that η has spin weight s. Then $P^{-s}\eta$ has spin weight
0 and so $\bar\eth(P^{-s}\eta) = P\partial(P^{-s}\eta)/\partial\zeta$, or $\bar\eth\eta = P^{1+s}\partial(P^{-s}\eta)/\partial\zeta$. We
write this equation and its complex conjugate as

$$\eth\eta = P^{1-s}\frac{\partial}{\partial\bar\zeta}(\eta P^s), \qquad \bar\eth\eta = P^{1+s}\frac{\partial}{\partial\zeta}(\eta P^{-s}). \tag{3.9.18}$$

These are often more useful than the original representation
(3.9.7), and indeed can be used to generalize $\eth$ to act on objects of
non-integer spin weight. (See exercise 3.9.5 below.) It is important
to realise that choosing ζ rather than $\ln\zeta$ as our holomorphic coordinate implies that the $\eth,\bar\eth$ defined by (3.9.18) are not the same as
those defined by (3.9.7). Although the $_0Y_{lm}$ defined by (3.9.9) and
(3.9.19) below are the same, the spin-weighted spherical harmonics
also differ. **The reader has been warned!**

(3.9.5) Exercise. Suppose that A, B have spin weights a, b where
$a + b = -1$. Show that

$$\oint A(\eth B)\,dS = -\oint B(\eth A)\,dS,$$

where $dS = P^{-2}d\xi d\bar\xi$.

(3.9.6) Exercise. Verify the commutator relation (3.9.8) using the
representation (3.9.18).

A representation of the standard spherical harmonics is given
by

$$\begin{aligned}_0Y_{lm} = (-)^m &\left(\frac{2l+1}{4\pi}\right)^{1/2} \frac{l!\,[(l+m)!(l-m)!]^{1/2}}{(1+\zeta\bar\zeta)^l} \times \\ &\sum_n \frac{(-)^n\zeta^{l-n}\bar\zeta^{l-m-n}}{n!(l-m-n)!(l-n)!(m+n)!}.\end{aligned} \tag{3.9.19}$$

where the sum is from $\max(0, -m)$ to $\min(l, l - m)$. The spin-weighted spherical harmonics become, in this representation

$$_sY_{lm} = (-)^m \left(\frac{2l+1}{4\pi}\right)^{1/2} \frac{[(l+s)!(l-s)!(l+m)!(l-m)!]^{1/2}}{(1+\zeta\bar{\zeta})^l} \times$$

$$\sum_n \frac{(-)^n \zeta^{l+s-n} \bar{\zeta}^{l-m-n}}{n!(l-m-n)!(l+s-n)!(m+n-s)!}$$

(3.9.20)

where the sum is from $\max(0, s - m)$ to $\min(l + s, l - m)$.

(3.9.7) Problem. Compute the four independent quantities $_{\pm\frac{1}{2}}Y_{\frac{1}{2}\pm\frac{1}{2}}$. Consider the set S of quantities of the form $a\bar{b}$, where each of a and b is one of the above harmonics. Of what group is S a representation?

The operators $\eth, \bar{\eth}$ and the spin-weighted spherical harmonics are explored in some detail in the article by Goldberg et al., 1967. However they consider a sphere of radius $2^{-1/2}$ with a positive definite signature and inward normal. Thus their operator, denoted $\eth_G$, satisfies

$$\eth_G = -2^{1/2}R\bar{\eth}.$$

The sign change comes from the signature, and the factor from the change in radius. Since they regard ζ rather than $\bar{\zeta}$ as holomorphic they have effectively interchanged m^a and $\bar{m}^a$, and this causes the complex conjugation. It also means that

$$s_G = -s.$$

The spin-weighted spherical harmonics defined here are the same as those of Penrose and Rindler, 1984. It should be noted however that although a number of earlier papers made use of the harmonics and $\eth$ there appears to be no consistent notation.

3.10   Asymptotic solution of the field equations

In the later sections we shall need properties of the space-time in a neighbourhood of $\mathscr{I}$. The information obtained in section 3.7

applies only on $\mathscr{I}$. We must therefore develop asymptotic expansions for the NP scalars. This was first done by Newman and Unti, 1962, and most of their successors have adopted the same technique. With a little foresight and a careful choice of coordinates and tetrad however much of the tedious algebra can be avoided. Because almost all of our calculations take place in the physical space-time we remove the tilde from physical quantities and attach a caret to unphysical quantities, e.g., (3.1.5).

Let (M, g) be an asymptotically simple space-time. Suppose there is a 1-parameter family of null hypersurfaces N_u intersecting $\mathscr{I}$ in cuts S_u. We may use the parameter to define a scalar field u in M via $u = const.$ on N_u. On each geodesic generator γ_u of N_u we choose an affine parameter r. On any one cut of $\mathscr{I}$, say S_0 we choose arbitrary coordinates (x^α), $\alpha, \beta, \ldots = 2, 3$. We propagate x^α onto $\mathscr{I}$ by requiring $x^\alpha = const.$ on the generators of $\mathscr{I}$. Finally we propagate x^α into M via $x^\alpha = const.$ on each γ_u. Now (u, r, x^α) is a coordinate chart in a neighbourhood of $\mathscr{I}$. The remaining freedom is

1) $u \mapsto \hat{u} = \hat{u}(u)$, a relabelling of the null hypersurfaces,
2) $x^\alpha \mapsto \hat{x}^\alpha = \hat{x}^\alpha(x^\beta)$, a different choice of x^α on S_0,
3) $r \mapsto \hat{r} = a(u, x^\alpha)r + b(u, x^\alpha)$, the scaling and origin for r on each generator γ_u.

We next set up the NP tetrad. Let

$$l_a = \nabla_a u. \qquad (3.10.1)$$

Then l^a is an affinely parametrized geodesic vector, and so we may use the scaling freedom in r to set

$$l^a \nabla_a = \partial/\partial r. \qquad (3.10.2)$$

We next label the 2-surfaces $u = const.$, $r = const.$ as S_{ur}. At each point p on S_{ur} l^a is future pointing and orthogonal to S_{ur}. There is precisely one other null direction at p with this property, and we choose n^a to be parallel to it. Finally we choose $m^a, \overline{m}^a$ to span $T(S_{ur})$. The freedom remaining is

1) a *boost* $l \mapsto Al$, $n \mapsto A^{-1}n$, which allows us to rescale r again,

2) a **spin** $m \mapsto e^{i\theta}m$, with θ real.

From the construction we deduce that

$$n^a \nabla_a = \partial/\partial u + Q\partial/\partial r + C^\alpha \partial/\partial x^\alpha, \quad m^a \nabla_a = P^\alpha \partial/\partial x^\alpha,$$
$$(3.10.3)$$

where Q, C^α, P^α represent 6 real functions, the metric components. (We may always recover the metric from $g_{ab} = 2l_{(a}n_{b)} - 2m_{(a}\overline{m}_{b)}$.) Further $C^\alpha = 0$ on $\mathscr{I}$. (NB. Since this is not the Newman-Unti tetrad, the notation has been changed.)

Under a spin, $\epsilon - \bar{\epsilon} = m^a D\overline{m}_a \mapsto \epsilon - \bar{\epsilon} + iD\theta$. We shall choose $D\theta$ to set $\epsilon = \bar{\epsilon}$ everywhere. However θ can still be chosen arbitrarily on any one surface $r = const$. Now applying the commutators to x^a gives

$$\kappa = \epsilon = 0, \quad \tau = \bar{\pi} = \bar{\alpha} + \beta, \quad \rho \text{ and } \mu \text{ are real}, \quad (3.10.4)$$

and the "frame equations"

$$DQ = -\gamma - \bar{\gamma}, \qquad (Fa)$$

$$DC^\alpha = 2\bar{\tau}P^\alpha + 2\tau\overline{P}^\alpha, \qquad (Fb)$$

$$DP^\alpha = \rho P^\alpha + \sigma\overline{P}^\alpha, \qquad (Fc)$$

$$\Delta P^\alpha - \delta C^\alpha = (\gamma - \bar{\gamma} - \mu)P^\alpha - \bar{\lambda}\overline{P}^\alpha, \qquad (Fd)$$

$$\delta\overline{P}^\alpha - \bar{\delta}P^\alpha = (\bar{\beta} - \alpha)P^\alpha - (\beta - \bar{\alpha})\overline{P}^\alpha, \qquad (Fe)$$

$$\delta Q = -\bar{\nu}. \qquad (Ff)$$

For this and the next sections we are denoting the "unphysical" quantities by a hat. It is easy to see from (3.5.8), (3.4.5) that $\Psi_n = \Omega^{4-n}\widehat{\Psi}_n$. Since we know that near $\mathscr{I}$, $\widehat{\Psi}_n = O(\Omega)$ we deduce that

$$\Psi_n = O(\Omega^{5-n}). \qquad (3.10.5)$$

Using the definitions, the rules of section 3.4 and (3.5.8) it is straightforward to show, assuming the NP scalars to be regular near $\mathscr{I}$, that σ is $O(\Omega^2)$, τ, α, β, π are $O(\Omega)$, γ, μ, λ are $O(1)$, ν is $O(\Omega^{-1})$ and

$$\rho = \Omega^{-1}D\Omega + O(\Omega^2).$$

We choose $D\Omega = d\Omega/dr = -\Omega^2$ so that $\rho = -\Omega + O(\Omega^2)$. Now we have not yet fixed the origin of the affine parameter r, and writing ρ in terms of r it is easy to see that we may choose this origin to set $\rho = -\Omega + O(\Omega^3)$. We next set $\sigma = s\Omega^2 + O(\Omega^3)$ where $s = s(u, x^\alpha)$ is Ω-independent. Substituting these expressions into the field equations (a, b) of appendix B and noting that $D = -\Omega^2 d/d\Omega$ we obtain

$$\rho = -\Omega - s\bar{s}\Omega^3 + O(\Omega^5), \qquad \sigma = s\Omega^2 + O(\Omega^4). \qquad (3.10.6)$$

We expect m^a and hence P^α to be $O(\Omega)$. Setting $P^\alpha = p^\alpha(u, x^\alpha)\Omega + O(\Omega^2)$ we find immediately from equation (Fc) that

$$P^\alpha = p^\alpha\Omega - s\bar{p}^\alpha\Omega^2 + O(\Omega^3). \qquad (3.10.7)$$

If we set $\delta^0 = p^\alpha\nabla_\alpha$ this can be written as

$$\delta = \Omega\delta^0 - s\Omega^2\bar{\delta}^0 + O(\Omega^3). \qquad (3.10.8)$$

Writing $\alpha = a\Omega + a^1\Omega^2 + O(\Omega^3)$, with a similar expansion for β field equations (d, e) reveal

$$\beta = -\bar{a}\Omega - as\Omega^2 + O(\Omega^3). \qquad (3.10.9)$$

The leading terms in equation (Fe) then show that

$$\delta^0\bar{p}^\alpha - \bar{\delta}^0 p^\alpha = -2ap^\alpha + 2\bar{a}\bar{p}^\alpha. \qquad (3.10.10)$$

Given a choice for p^α we can determine a. We now define $\eth$ acting on a spin s scalar η by

$$\eth\eta = \delta^0\eta + 2s\bar{a}\eta, \qquad \bar{\eth}\eta = \bar{\delta}^0\eta - 2sa\eta, \qquad (3.10.11)$$

consistent with section 3.9. The leading terms in field equation (k) now give a^1 and so

$$\alpha = a\Omega + (\eth\bar{s} + \bar{a}\bar{s})\Omega^2 + \Omega(\Omega^3). \qquad (3.10.12)$$

Since $\tau = \bar{\alpha} + \beta$ we have

$$\bar{\pi} = \tau = (\bar{\eth}s)\Omega^2 + O(\Omega^3). \qquad (3.10.13)$$

Also the Bianchi identity (Ba) plus the estimates (3.10.5) imply

$$\Psi_0 = \Psi_0^0\Omega^5 + O(\Omega^6), \qquad \Psi_1 = \Psi_1^0\Omega^4 - (\bar{\partial}\Psi_0^0)\Omega^5 + O(\Omega^6),$$

$$(3.10.14)$$

where the superfix denotes an Ω-independent quantity. We might have expected C^α to be $O(1)$, but our coordinates and frame were chosen so that C^α vanished on $\mathscr{I}$. Assuming $C^\alpha = O(\Omega)$ equation (Fb) reveals

$$C^\alpha\nabla_\alpha = -\left[(\partial\bar{s})\delta^0 + (\bar{\partial}s)\bar{\delta}^0\right]\Omega^2 + O(\Omega^3). \qquad (3.10.15)$$

The next group of equations are a little trickier to get started. The Bianchi identity (Bc) plus the estimates (3.10.5) reveal

$$\Psi_2 = \Psi_2^0\Omega^3 - (\bar{\partial}\Psi_1^0)\Omega^4 + O(\Omega^5). \qquad (3.10.16)$$

We expect $\gamma = O(1)$ and field equation (f) reveals

$$\gamma = \gamma^0 - \left[a\bar{\partial}s - \bar{a}\partial\bar{s} + \tfrac{1}{2}\Psi_2^0\right]\Omega^2 + O(\Omega^3).$$

Equation (Fa) then gives

$$Q = -(\gamma^0 + \bar{\gamma}^0)/\Omega + Q^0 - \tfrac{1}{2}(\Psi_2^0 + \overline{\Psi}_2^0)\Omega + O(\Omega^2).$$

Next we set

$$\lambda = \lambda^0 + \lambda^1\Omega + \lambda^2\Omega^2 + O(\Omega^3),$$

with a similar expression for μ. Equations (g,h) immediately give $\lambda^0 = \mu^0 = 0$. But now the leading, $O(\Omega)$ terms in equation (Fd) reveal that $\bar{\gamma}^0 = -\tfrac{1}{2}\Delta p^\alpha/p^\alpha$ where $\Delta = \partial/\partial u$. Now p^α is free data on $\mathscr{I}$ and we shall always choose it to set $\Delta p^\alpha = 0$. Thus we may conclude that $\gamma^0 = 0$, and

$$\gamma = -\left[a\bar{\partial}s - \bar{a}\partial\bar{s} + \tfrac{1}{2}\Psi_2^0\right]\Omega^2 + O(\Omega^3). \qquad (3.10.17)$$

The leading $O(\Omega^2)$ terms in (p) give $\lambda^1 = \Delta\bar{s}$. The leading $O(\Omega^2)$ terms in (l) give

$$\mu^1 = m = -\delta^0 a - \bar{\delta}^0\bar{a} + 4a\bar{a}. \qquad (3.10.18)$$

Next (g,h) give $\lambda^2 = -m\bar{s} - \bar{\eth}\eth\bar{s}$, $\mu^2 = -\eth^2\bar{s} - s\Delta\bar{s} - \Psi_2^0$. Finally the $O(\Omega^2)$ terms in (q) give $Q^0 = m$. Thus

$$\lambda = (\Delta\bar{s})\Omega - [m\bar{s} + \bar{\eth}\eth\bar{s}]\Omega^2 + O(\Omega^3), \qquad (3.10.19)$$

$$\mu = m\Omega - [\eth^2\bar{s} + s\Delta\bar{s} + \Psi_2^0]\Omega^2 + O(\Omega^3), \qquad (3.10.20)$$

$$Q = m - \tfrac{1}{2}(\Psi_2^0 + \overline{\Psi}_2^0)\Omega + O(\Omega^2). \qquad (3.10.21)$$

The final group of equations is straightforward. The Bianchi identities (Be,g) reveal

$$\Psi_3 = \Psi_3^0\Omega^2 - (\bar{\eth}\Psi_2^0)\Omega^3 + O(\Omega^4), \qquad (3.10.22)$$

$$\Psi_4 = \Psi_4^0\Omega - (\bar{\eth}\Psi_3^0)\Omega^2 + O(\Omega^3), \qquad (3.10.23)$$

and equation (Ff) gives

$$\nu = -(\bar{\eth}m)\Omega + \left[\tfrac{1}{2}\bar{\eth}(\Psi_2^0 + \overline{\Psi}_2^0) + \bar{s}\eth m\right]\Omega^2 + O(\Omega^3). \qquad (3.10.24)$$

There are some more useful equations relating the leading terms in the Weyl scalars. Since the Ψ_n^0 are precisely the χ_n of section 3.7 and the s here is the σ there we can read some of them off directly from (3.7.14–16) as

$$\Psi_4^0 = -\Delta^2\bar{s}, \qquad \Psi_3^0 = -\eth\Delta\bar{s}, \qquad (3.10.25)$$

$$\Im[\Psi_2^0 + s\Delta\bar{s} + \eth^2\bar{s}] = 0. \qquad (3.10.26)$$

We also need one other result, obtained from the leading order terms in equation (Bf), viz.

$$\Delta\Psi_2^0 = \eth\Psi_3^0 + s\Psi_4^0. \qquad (3.10.27)$$

If higher order terms are required it is very easy to generate them once the lower order terms are known, provided the equations are handled in the order indicated.

For use in the next section we construct the a and m appropriate to spherical cuts of $\mathscr{I}$. Following sections 3.8 and 3.9 we set

$$P^{02} = 2^{-1/2}, \qquad P^{03} = -2^{-1/2}i\operatorname{cosec}\theta. \qquad (3.10.28)$$

Then equation (3.10.10) reveals

$$a = -2^{-3/2} \cot \theta, \qquad (3.10.29)$$

and (3.10.18) then gives

$$m = -\tfrac{1}{2}.$$

3.11   The Bondi 4-momentum covector

In this, the penultimate section of the chapter, a 4-(co-)vector P_a will be defined with the following properties:

a) P_a is defined for each cut of $\mathscr{I}$, and if t^a is a future-pointing timelike vector field at $\mathscr{I}$ then P_a defines a scalar $M_B = P_a t^a$,

b) if t^a is chosen as the timelike Killing vector for the Kerr family of spacetimes then M_B is the mass parameter,

c) $M_B \geq 0$ provided the dominant energy condition holds,

d) if $M_B(u)$ is defined for a Bondi system of cuts then $\Delta M_B \leq 0$.

Thus the **Bondi mass** M_B is to be interpreted as the mass of an isolated system. It is positive or zero, identifiable with the mass in stationary situations and decreases as the system radiates. Thus an isolated system can only radiate a finite amount of energy. There is no possibility of extracting an infinite amount of energy by allowing the "Newtonian potential energy" to decrease indefinitely. In this section only property a) is verified. The remainder are dealt with in section 3.12. Note that throughout this section a quantity in the physical space-time will not carry a tilde since most of the calculations will be done there. Instead quantities in the unphysical space-time will be denoted by a caret, e.g., equation (3.1.5).

We shall use the asymptotic null tetrad developed in section 3.10. Since the first part of the analysis occurs in the unphysical space-time we need to be able to translate the results. It will be recalled from equation (3.5.8) that we are adopting the conformal transformation formulae

$$\hat{o}_A = o_A, \quad \hat{\iota}_A = \Omega \iota_A, \quad \hat{o}^A = \Omega^{-1} o^A, \quad \hat{\iota}^A = \iota^A. \qquad (3.11.1)$$

This implies for the frame components (3.10.3) that

$$\widehat{Q} = \Omega^{-2}Q, \quad \widehat{C}^\alpha = C^\alpha, \quad \widehat{P}^\alpha = \Omega^{-1}P^\alpha, \tag{3.11.2}$$

and that $\widehat{D} = d/d\Omega$. To determine the NP scalars we need to use the formulae of section 3.4, obtaining

$$\begin{aligned}
\widehat{\kappa} &= \Omega^{-3}\kappa, & \widehat{\epsilon} &= \Omega^{-2}\epsilon, & \widehat{\pi} &= \Omega^{-1}(\pi + \bar{\delta}\Upsilon), \\
\widehat{\rho} &= \Omega^{-2}(\rho - D\Upsilon), & \widehat{\alpha} &= \Omega^{-1}(\alpha - \bar{\delta}\Upsilon), & \widehat{\lambda} &= \lambda, \\
\widehat{\sigma} &= \Omega^{-2}\sigma, & \widehat{\beta} &= \Omega^{-1}\beta, & \widehat{\mu} &= \mu + \Delta\Upsilon, \\
\widehat{\tau} &= \Omega^{-1}(\tau - \delta\Upsilon), & \widehat{\gamma} &= \gamma - \Delta\Upsilon, & \widehat{\nu} &= \Omega\nu,
\end{aligned}$$
$$\tag{3.11.3}$$

where $\Upsilon = \Omega^{-1}D\Omega$. It should be noted that on $\mathscr{I}$ all of these scalars vanish except for $\sigma = s$, $\alpha = a$, $\beta = -a$ in the notation of section 3.10. Finally

$$\widehat{\Psi}_n = \Omega^{n-5}\Psi_n = \Psi_n^0. \tag{3.11.4}$$

We now need a concept which has caused some confusion in the past. We would like a constant spinor, but as we saw in section 2.9 this is only possible in type N spacetimes. From the peeling theorem we know that an asymptotically simple spacetime is asymptotically of type N, and so we should be able to construct an asymptotically constant spinor.

(3.11.1) DEFINITION

*A spinor τ^A will be said to be **strongly asymptotically constant** in $\check{M}$ if*

i) $\tau_A o^A$, $\tau_A \iota^A$ *are regular at $\mathscr{I}$,*
ii) $\widehat{\nabla}^{A'}{}_{(A}\widehat{\tau}_{B)} = 0$ *at $\mathscr{I}$, where the conformal behaviour of τ is defined by $\tau^A = \widehat{\tau}^A$.*

The equation above is the **twistor equation**, and with the indicated conformal behaviour it is conformally invariant. If imposed generally it has non-trivial solutions only in type N spacetimes, but it can be imposed on certain subspaces, in particular on $\mathscr{I}$, and in this context it is known as the **asymptotic twistor equation**.

(3.11.2) LEMMA

Let

$$\hat{\tau}^A = \hat{x}\hat{o}^A + \hat{y}\hat{\imath}^A = xo^A + y\iota^A.$$

Then x, y are regular on $\mathscr{I}$ and if their values there are denoted x^0, y^0 then

$$\hat{x} = \Omega x^0 + \Omega^2 x^1 + O(\Omega^2), \qquad \hat{y} = y^0 + O(\Omega^2),$$

where the coefficients are Ω-independent and

$$(\hat{\delta} - \hat{\beta})y^0 = 0, \qquad (\hat{\bar{\delta}} - \hat{\alpha})y^0 = -x^0, \qquad \hat{\Delta}y^0 = 0 \qquad on \ \mathscr{I}.$$
$$(3.11.5)$$

Proof: $\tau^A o_A = \hat{\tau}^A \hat{o}_A = y$ is regular and so we may set $y = y^0 + \Omega y^1 + O(\Omega^2)$. Similarly $\tau^A \iota_A = \Omega^{-1}\hat{\tau}^A \hat{\imath}_A = \Omega^{-1}\hat{x}$ is regular, and so we may set $\hat{x} = \Omega x^0 + \Omega^2 x^1 + O(\Omega^3)$. Projecting out the asymptotic equation into components on $\mathscr{I}$ now gives $y^1 = 0$ and (3.11.5).

With any univalent spinor τ we may associate a real null vector $t^a = \tau^A \bar{\tau}^{A'}$, a symmetric two spinor

$$\phi_{AB} = \tfrac{1}{2}\left(\tau_{(A}\nabla^{C'}{}_{B)}\bar{\tau}_{C'} - \bar{\tau}_{C'}\nabla^{C'}{}_{(A}\tau_{B)}\right), \qquad (3.11.6)$$

and a real bivector (skew symmetric valence 2 tensor)

$$F_{ab} = \epsilon_{A'B'}\phi_{AB} + \epsilon_{AB}\bar{\phi}_{A'B'}. \qquad (3.11.7)$$

We can now proceed to the construction of the Bondi 4-momentum. Let Σ be a null hypersurface in M which extends to $\mathscr{I}$, and let $S(\Omega)$ be the 2-surface $\Omega = const.$ in Σ. We now choose a null tetrad as in the previous section. l^a is tangent to the generators of Σ, n^a is orthogonal to the $S(\Omega)$ while $m^a, \bar{m}^a$ span the tangent space to $S(\Omega)$. Let

$$I(\Omega) = \oint_{S(\Omega)} F_{ab}l^a n^b \, dS \qquad (3.11.8)$$

where

$$\oint_{S(\Omega)} dS = \Omega^{-2}.$$

The Bondi 4-momentum now appears as the limit as $\Omega \to 0$.

(3.11.3) THEOREM

*Suppose that τ^A is strongly asymptotically constant. Then $I = \lim_{\Omega \to 0} I(\Omega)$ exists and defines the **generalized Bondi 4-momentum** P^a via $I = P^a t_a$ on $\mathscr{I}$.*

Proof: Set

$$p = \iota^B \delta \tau_B, \quad q = o^B \delta \tau_B, \quad u = o^B \bar{\delta} \tau_B,$$
$$v = o^B \Delta \tau_B, \quad w = \iota^B D \tau_B. \qquad (3.11.9)$$

Then a tedious but straightforward calculation using (3.11.5–7) shows that

$$F_{ab} l^a n^b = \Re[\bar{x} u + \bar{y} p]. \qquad (3.11.10)$$

Now

$$u = -(\bar{\delta} - \alpha) y + \rho x, \quad p = (\delta + \beta) x + \mu y.$$

Thus

$$F_{ab} l^a n^b = \Re \left[-\bar{x}(\bar{\delta} - \alpha) y + \bar{y}(\delta + \beta) x + \rho x \bar{x} + \mu y \bar{y} \right] \quad (3.11.11)$$

Since $dS = O(r^2) = O(\Omega^{-2})$ we have to show that the integrand (3.11.11) is $O(\Omega^2)$ in order for $I(\Omega)$ to converge as $\Omega \to 0$. Now x has spin $-\frac{1}{2}$ while y has spin $\frac{1}{2}$. If $\eth$ is defined for the unit sphere then the $O(\Omega)$ terms in (3.11.11) are found after making the relevant substitutions to be

$$
\begin{aligned}
O(\Omega)\text{terms} &= \tfrac{1}{2} \left[-x^0 \bar{x}^0 - \tfrac{1}{2} y^0 \bar{y}^0 - \bar{x}^0 \bar{\eth} y^0 + \bar{y}^0 \eth x^0 + c.c. \right] \\
&= \tfrac{1}{2} \left[-\bar{x}^0 (\bar{\eth} y^0 + x^0) - \tfrac{1}{2} y^0 \bar{y}^0 - \bar{y}^0 \eth \bar{\eth} y^0 + c.c. \right] \\
&= \tfrac{1}{2} \left[-\tfrac{1}{2} y^0 \bar{y}^0 - \bar{y}^0 (\bar{\eth} \eth y^0 - \tfrac{1}{2} y^0) + c.c. \right] \\
&= 0,
\end{aligned}
$$

where the commutator of $\eth, \bar{\eth}$ has been used together with (3.11.5) again. The $O(\Omega^2)$ terms are

$$O(\Omega^2)\text{terms} = \tfrac{1}{2}[-x^0\bar{x}^1 - \bar{x}^0 x^1 + \mu^1 y^0\bar{y}^0 - \bar{x}^1\eth y^0 + \bar{y}^0\eth x^1 +$$
$$\bar{x}^0\bar{s}\eth y^0 + (\eth\bar{s})\bar{x}^0 y^0 + \bar{y}^0(-s\eth)x^0 + c.c.].$$

$$(3.11.12)$$

The first and fourth terms cancel each other out and the sixth term vanishes because of (3.11.5). Next note that the second and fifth terms become, using (3.11.5)

$$-\bar{x}^0 x^1 + \bar{y}^0\eth x^1 = x^1\eth\bar{y}^0 + \bar{y}^0\eth x^1.$$

Now $spin(x) + spin(\bar{y}) = -1$. Using exercise 3.9.5 these terms will cancel on integration. The seventh term is $\bar{x}^0 y^0(\eth\bar{s})$. Since s has spin 2, $\bar{s}$ has spin -2. Thus using exercise 3.9.5, under integration this term delivers the same value as $-\bar{s}\eth(y^0\eth\bar{y}^0)$, or $\bar{s}y^0\eth^2\bar{y}^0$. Now y^0 has spin $\tfrac{1}{2}$ and can be expanded in terms of spin-weighted spherical harmonics

$$y^0 = \sum_{l,m} c_{lm}{}_{\frac{1}{2}}Y_{l+\frac{1}{2}\ m+\frac{1}{2}}(\theta,\phi),$$

where l ranges from 0 to ∞ and m ranges from $-(l+1)$ to l. However (3.11.5) implies $\eth y^0 = 0$ or

$$0 = \sum_{l,m} c_{lm} \times \text{irrelevant factor} \times {}_{\frac{3}{2}}Y_{l+\frac{1}{2}\ m+\frac{1}{2}}(\theta,\phi),$$

This means that $c_{lm} = 0$ for $l > 0$, since the spherical harmonics are independent functions. We can say nothing about the $l = 0$ coefficients because the corresponding spherical harmonics vanish. Now

$$\bar{\eth}^2 y^0 = \sum_{m} c_{0m} \times \text{irrelevant factor} \times {}_{-\frac{3}{2}}Y_{\frac{1}{2}\ m+\frac{1}{2}}(\theta,\phi),$$

vanishes because ${}_{-s}Y_{lm} = 0$ for $s > l$. Thus the seventh term in (3.11.12) will vanish on integration. By a similar argument the eighth term also vanishes on integration. All that remains is the third term and its complex conjugate, which because of equation (3.10.26) contribute

$$-\tfrac{1}{2}y^0\bar{y}^0[\Psi_2^0 + s\Delta\bar{s} + \eth^2\bar{s}].$$

The last term here vanishes also on integration for precisely the reasons given above. Thus

$$I(0) = -\tfrac{1}{2}\oint_{S(0)} y^0\bar{y}^0[\Psi_2^0 + s\Delta\bar{s}]\,dS, \qquad (3.11.13)$$

exists and is finite.

It is clear that strong asymptotic constancy requires

$$y^0 = c_{+\frac{1}{2}}Y_{\frac{1}{2}\frac{1}{2}} + c_{-\frac{1}{2}}Y_{\frac{1}{2}-\frac{1}{2}}, \qquad (3.11.14)$$

where the coefficients $c_\pm$ are constants. Now problem 3.9.7 showed that, ignoring irrelevant constant factors

$$y^0\bar{y}^0 = \frac{A + B\zeta + \bar{B}\bar{\zeta} + C\zeta\bar{\zeta}}{1 + \zeta\bar{\zeta}} \qquad (3.11.15)$$

where the constants A, C are real and B is complex. But comparison with equation (3.8.7) shows that the freedom in $y^0\bar{y}^0$ is precisely that of the translation subgroup of the BMS group. Note that the 4 vectors t^a implied by independent choices of the constants in (3.11.15) are all parallel to n^a on $\mathscr{I}$ and so are not independent as space-time vectors. However they are independent vectors when regarded as elements of the Lie algebra of the BMS group. Choosing the constants in (3.11.14) to generate the element corresponding to an x-translation generates the x-component of P_a via (3.11.13). In this sense (3.11.11) furnishes the Bondi 4-momentum at $\mathscr{I}$.

3.12   The positivity of the Bondi mass

In this final section of the chapter, we show that the Bondi mass is positive, i.e., $P_a t^a \geq 0$ for all future pointing timelike vectors t^a. We need a preliminary lemma. With any univalent spinor τ

we have associated a real null vector $t^a = \tau^A \bar{\tau}^{A'}$, a symmetric two spinor

$$\phi_{AB} = \tfrac{1}{2}\left(\tau_{(A}\nabla_{B)}{}^{C'}\bar{\tau}_{C'} - \bar{\tau}_{C'}\nabla^{C'}{}_{(A}\tau_{B)}\right), \qquad (3.12.1)$$

and a real bivector

$$F_{ab} = \epsilon_{A'B'}\phi_{AB} + \epsilon_{AB}\bar{\phi}_{A'B'}. \qquad (3.12.2)$$

The divergence of F_{ab} will be needed several times, and so we compute it now.

$(3.12.1)$ LEMMA

$$\nabla^b F_{ab} = \tfrac{1}{2}((\nabla_{A'}{}^B \bar{\tau}_{C'})(\nabla^{C'}{}_{(A}\tau_{B)}) - (\nabla_{A'}{}^B \tau_{(A})(\nabla_{B)}{}^{C'}\bar{\tau}_{C'}) + c.c.)$$
$$- \tfrac{1}{2}G_{ab}t^b. \qquad (3.12.3)$$

Proof: Unfortunately this calculation is somewhat messy. For the sake of brevity only one half of the terms are shown explicitly, although all are needed to see some of the cancellations that occur. Clearly

$$4\nabla^b F_{ab} = 4\epsilon_{A'B'}\nabla^{BB'}\phi_{AB} + c.c. = -4\nabla_{A'}{}^B \phi_{AB} + c.c.$$
$$= 2((\nabla_{A'}{}^B \bar{\tau}_{C'})(\nabla^{C'}{}_{(A}\tau_{B)}) -$$
$$(\nabla_{A'}{}^B \tau_{(A})(\nabla_{B)}{}^{C'}\bar{\tau}_{C'}) + c.c.) + g_a,$$

where

$$g_a = \bar{\tau}^{C'}\nabla_{A'B}\nabla_{C'A}\tau^B + \bar{\tau}^{C'}\nabla_{A'B}\nabla_{C'}{}^B \tau_A -$$
$$\tau_A \nabla_{A'B}\nabla^B{}_{C'}\bar{\tau}^{C'} - \tau^B \nabla_{A'B}\nabla_{AC'}\bar{\tau}^{C'} + c.c.$$

$$\begin{aligned}
&= \bar{\tau}^{C'} \nabla_{C'A} \nabla_{A'B} \tau^B + \bar{\tau}^{C'} \epsilon_{A'C'} \Box_{BA} \tau^B + \\
&\quad \bar{\tau}^{C'} \epsilon_{BA} \Box_{A'C'} \tau^B + \bar{\tau}^{C'} \Box_{A'C'} \tau_A + \\
&\quad \tfrac{1}{2} \bar{\tau}^{C'} \epsilon_{A'C'} \Box \tau_A - \tau_A \Box_{A'C'} \bar{\tau}^{C'} - \tfrac{1}{2} \tau_A \epsilon_{A'C'} \Box \bar{\tau}^{C'} - \\
&\quad \tau^B \nabla_{A'B} \nabla_{AC'} \bar{\tau}^{C'} + c.c. \\
&= -\bar{\tau}_{A'} \Box_{AB} \tau^B + 2 \bar{\tau}^{C'} \Box_{A'C'} \tau_A - \tfrac{1}{2} \bar{\tau}_{A'} \Box \tau_A - \\
&\quad \tau_A \Box_{A'C'} \bar{\tau}^{C'} + \tfrac{1}{2} \tau_A \Box \bar{\tau}_{A'} + c.c. \\
&= -2 \bar{\tau}_{A'} \Box_{AB} \tau^B + 2 \bar{\tau}^{C'} \Box_{A'C'} \tau_A - + c.c. \\
&= -2 \bar{\tau}_{A'} (-3 \Lambda \tau_A) + 2 \bar{\tau}^{C'} \Box_{A'C'} \tau_A + c.c. \\
&= +4 \Phi_{ab} t^b + 12 \Lambda t_a \\
&= -2 G_{ab},
\end{aligned}$$

using Einstein's field equations.

In section 3.11 we constructed a null hypersurface Σ which extended to $\mathscr{I}$, having affinely parametrized generators with tangent vector l^a. Σ was foliated by two-surfaces $S(\Omega)$ on which Ω was constant. On each $S(\Omega)$ we defined an integral

$$I(\Omega) = \oint_{S(\Omega)} F_{ab} l^a n^b \, dS,$$

where F_{ab} was constructed from a univalent spinor τ. If τ was strongly asymptotically constant we saw that $I(0)$ furnished the Bondi 4-momentum at $\mathscr{I}$. We may propagate τ from $\mathscr{I}$ into Σ in an arbitrary manner, and as we show below this can be done so that $I(\Omega)$ is a monotone non-increasing function of Ω. We need however a definition.

(3.12.2) DEFINITION

*An energy-momentum tensor T^{ab} is said to obey the **dominant energy condition** if it obeys the weak energy condition and in addition $T^{ab} v_b$ is a future-pointing non-spacelike vector for all future-pointing timelike vectors v^a.*

Physically this means that in any orthonormal frame the energy density μ dominates other terms

$$T^{00} \geq T^{ab} \qquad \textit{for all } a, b.$$

This is satisfied for all known forms of matter; if it were not then the speed of sound $\sqrt{(dp/d\mu)}$ could exceed 1 and causality would be lost. It follows immediately from the definition that

$$T_{ab} w^a v^b \geq 0,$$

for all timelike v^a and non-spacelike w^a, both vectors being future-pointing.

(3.12.3) THEOREM

Suppose that τ^A is propagated on Σ by

$$D\tau^A = uo^A, \qquad\qquad (3.12.4)$$

where $u = o^B \bar{\delta} \tau_B$, and the dominant energy condition holds. Then $dI(\Omega)/d\Omega \leq 0$.

Proof: Suppose we truncate Σ by $S_1 = S(\Omega_1)$ and $S_2 = S(\Omega_2)$, where $\Omega_1 \geq \Omega_2$. The divergence theorem implies

$$I(\Omega_2) - I(\Omega_1) = \oint_{\Sigma} (\nabla^b F_{ab}) l^a \, d\Sigma.$$

Now from lemma (3.12.1)

$$l^a \nabla^b F_{ab} = q\bar{q} + u\bar{u} - \tfrac{1}{2} G_{ab} l^a t^b,$$

where $q = o^B \delta \tau_B$, and (3.12.4) has been used. Now by Einstein's field equation $G_{ab} = -8\pi G T_{ab}$ and so $l^a \nabla^b F_{ab} \geq 0$, which is sufficient to prove the assertion.

We next consider a compact spacelike hypersurface A bounded by some $S(\Omega)$. This will have a timelike unit normal which we denote by v^a. v^a defines a projection operator

$$h_a{}^b = \delta_a{}^b - v_a v^b,$$

which annihilates vectors parallel to v. We can use h to define a derivative D_a intrinsic to A via

$$D_a = h_a{}^b \nabla_b = \nabla_a - v_a v^b \nabla_b. \qquad (3.12.5)$$

With these preliminaries we can now proceed to the next result.

(3.12.4) THEOREM

*Let A be a compact spacelike hypersurface with unit timelike normal v^a, and boundary $S(\Omega_1)$. Suppose we require the spinor τ^A to satisfy the **Witten equation***

$$D_a \tau^B = 0, \qquad (3.12.6)$$

in A. Then if the dominant energy condition holds $I(\Omega_1) \geq 0$.

Proof: The divergence theorem implies that

$$I(\Omega_1) = \oint_A v^a \nabla^b F_{ab} \, dA.$$

Now $\nabla_a = D_a + v_a v^b \nabla_b$. Setting $\gamma_B = v^c \nabla_c \tau_B$ we find $\nabla_a \tau_B = D_a \tau_B + v_a \gamma_B = v_a \gamma_B$ if the Witten equation holds. Using lemma 3.12.1 it is now easy to show that

$$v^a \nabla^b F_{ab} = -\tfrac{1}{2} G_{ab} t^a v^b,$$

and the result follows.

Following Ludvigsen and Vickers, 1982 we can now prove the positivity of the Bondi mass on a cut S of $\mathcal{I}$. On S we choose a strongly asymptotically constant spinor τ^A. It is clear from lemma 3.11.2 that this is equivalent to choosing a u-independent solution of $\eth y^0 = 0$, i.e., y^0 has to have the form (3.11.14). Let Σ be the other null hypersurface through S whose generators are orthogonal to $\mathcal{I}$. We propagate τ^A from S to Σ via equation (3.12.4). The surfaces $\Omega = constant$ foliate Σ and we choose a particular one $S_1 : \Omega = \Omega_1$. Let A be any compact spacelike hypersurface with boundary S_1. We now propagate τ^A from S_1 onto A via the Witten equation (3.12.6). This is a fairly unusual boundary value problem for an elliptic system. However existence, (but not uniqueness,) is guaranteed by the corollary to theorem 10.6.1 of Hörmander, 1976. Now theorem 3.12.4 asserts $I(\Omega_1) \geq 0$ provided the dominant energy condition holds. Theorem 3.12.3 now guarantees that $I(0) \geq I(\Omega_1) \geq 0$. If one could assert not only existence but also uniqueness of the boundary value problem (3.12.6) one would have a local energy-momentum flux construction.

It finally remains to show that the Bondi mass is non-increasing. Up to a positive factor we may construct M_B by taking $y^0 = 1$ in (3.11.11), corresponding to a timelike translation, and so

$$M_B = -\tfrac{1}{2} \oint (\Psi_2^0 + s\Delta\bar{s})\, dS. \qquad (3.12.7)$$

Thus

$$\Delta M_B = -\tfrac{1}{2} \oint [\Delta(\Psi_2^0 + s\Delta\bar{s})]\, dS$$

$$= -\tfrac{1}{2} \oint (\Delta\Psi_2^0 + (\Delta s)(\Delta\bar{s}) + s\Delta^2\bar{s})\, dS$$

$$= -\tfrac{1}{2} \oint (\partial\Psi_3^0 + s\Psi_4^0 + (\Delta s)(\Delta\bar{s}) + s\Delta^2\bar{s})\, dS$$

using (3.10.27),

$$= -\tfrac{1}{2} \oint (\partial\Psi_3^0 - s\Delta^2\bar{s} + (\Delta s)(\Delta\bar{s}) + s\Delta^2\bar{s})\, dS$$

using (3.10.25),

$$= -\tfrac{1}{2} \oint (\partial\Psi_3^0 + (\Delta s)(\Delta\bar{s}))\, dS.$$

Now Ψ_3^0 has spin -1 and so its contribution to the integral will vanish because of exercise 3.9.5. Thus

$$\Delta M_B = -\tfrac{1}{2} \oint \Delta s\Delta\bar{s}\, dS \leq 0. \qquad (3.12.8)$$

(3.12.5) Problem. Consider a spacetime with coordinates (u, r, θ, ϕ) where $-\infty < u < \infty$, $0 < r < \infty$ and θ and ϕ are coordinates on the unit sphere. A NP tetrad can be chosen so that

$$l^a = (0,1,0,0), \qquad n^a = (1, -\tfrac{1}{2}F, 0, 0),$$

$$m^a = -2^{1/2}r^{-1}(0,0,1,-i/\sin\theta),$$

where $F(r) = 1 - 2m/r$ and m is a constant. Evaluate the metric and the NP scalars. Is the spacetime (weakly) asymptotically simple, and if so what is its Bondi mass?

(3.12.6) Problem. In prespinoric days Komar, 1959, tried to construct the F_{ab} of this section as a quantity linear in the derivative of a covector k trying the choices

$$F_{ab} = \lambda \epsilon_{ab}{}^{cd} \nabla_c k_d, \qquad F_{ab} = \lambda \nabla_{[a} k_{b]},$$

where λ is a constant. By computing the "current" $J^a = \nabla_b F^{ab}$, or otherwise, show that one of these choices is totally useless. On the other hand if k is timelike and Killing, the other choice leads to a plausible expression for the "mass" within a spacelike hypersurface A bounded by a spacelike 2-surface Σ, which is Σ-independent provided that Σ lies in the vacuum region surrounding an isolated source.

4

THE CHARACTERISTIC
INITIAL VALUE PROBLEM

The first three sections of this chapter are designed mainly for physicists (and especially relativists) who have little knowledge of the theory of partial differential equations. Its secondary purpose is to set out the notation and definitions; the many divergences in the literature result in considerable and unnecessary difficulties. We shall assume throughout that there are n independent variables denoted x^a, where $a, b, \ldots = 1, 2, \ldots, n$. Since our considerations will be mainly local we shall ignore topological issues and regard x^a as denoting a point x in a differentiable manifold M which may be taken to be R^n. In accordance with relativists' conventions partial differentiation will be denoted by a comma, e.g., $f_{,a} = \partial f / \partial x^a$.

4.1 Quasilinear first order
hyperbolic systems of equations

We now turn to first order quasilinear systems. The N dependent variables will be denoted ψ_α where $\alpha\beta, \gamma, \ldots = 1, 2, \ldots, N$. The general first order system can be expressed as

$$F(\psi_{\alpha,a}, \psi_\beta, x^a) = 0, \qquad (4.1.1)$$

for some vector-valued function F.

(4.1.1) DEFINITION

If the system (4.1.1) is linear in the $\psi_{\alpha,a}$ and there are precisely N equations, so that the system can be written as

$$A^a_{\alpha\beta} \psi_{\beta,a} = B_\alpha, \qquad (4.1.2)$$

167

*where the $A^a_{\alpha\beta}, B_\alpha$ are functions of the x^a and ψ_a, then (4.1.2) is a **quasilinear system**. Note that there are two sets of indices and an implicit sum for each set in (4.1.2). Under general functional transformations of the x^a, $\psi_{\beta,a}$ transforms like a covector, and $A^a_{\alpha\beta}$ like a vector. Therefore the usual relativists' convention on the positioning of these indices has been retained. Under linear transformations of the dependent variables, ψ_β and $A^a_{\alpha\beta}$ transform affinely. We shall always write the indices $\alpha, \beta \ldots$ in the lower position; repeated indices imply a sum. If the $A^a_{\alpha\beta}$ are independent of the ψ_β and the B_α are linear functions of the dependent variables ψ_β then (4.1.2) is a **linear system** of equations.*

(4.1.2) DEFINITION

*For a given solution ψ_β of (4.1.1), if when a is held fixed $A_{\alpha\beta}$ is hermitian, i.e., $A^a_{\alpha\beta} = \overline{A}^a_{\beta\alpha}$ for each a, then we have a **symmetric system for this solution**. If in addition a covector field ξ_a can be found so that the matrix $A^a_{\alpha\beta}\xi_a$ is positive-definite the system is said to be a **hyperbolic system for this solution**. For linear systems symmetry and hyperbolicity do not depend on a particular solution.*

Most of the equations of mathematical physics can be written as first order quasilinear symmetric hyperbolic systems and we give a number of examples. None are investigated in detail; instead we shall concentrate on identifying common features by which the reader can decide whether any chosen theory can be described in such terms.

Our first example is the zero rest mass scalar wave equation on a given background. Here we have a (possibly complex) scalar field ϕ satisfying $\Box \phi = 0$. Let $\theta_{AA'} = \nabla_{AA'}\phi$ denote the first derivatives of ϕ. Then the relevant field equation is

$$\nabla^{AA'}\theta_{AA'} = 0. \tag{4.1.3}$$

Note that the principal part of (4.1.3) involves a spinor divergence. (It also involves a second contraction, which is non-generic.)

Our second example is electromagnetism on a given spacetime background. Here there are 6 dependent variables, either the 3-vectors usually called **E** and **B** or their encoding (chapter 2) into a 2-component symmetric spinor ϕ_{AB}. (ϕ_{AB} may be thought of as a complex 2×2 symmetric matrix so that the number of real variables is again 6.) The sourcefree field equations are (2.5.16), i.e.

$$\nabla^{AA'}\phi_{AB} = 0. \qquad (4.1.4)$$

Note that the principal part of (4.1.4) involves a spinor divergence.

Our third example is Yang-Mills theory on a given spacetime background. Here the relevant field equations are, Penrose and Rindler, 1984, equation (5.5.44)

$$\nabla^{BA'}\phi_{AB\Theta\Psi} = 0, \qquad \nabla^{AB'}\chi_{A'B'\Theta\Psi} = 0, \qquad (4.1.5)$$

where Θ, Ψ take values in the appropriate frame bundle.

Next we consider linearized vacuum gravitational perturbations of a given vacuum spacetime background. Here the dependent variables are the 10 independent components of the Weyl tensor encoded as the components of the totally symmetric Weyl spinor Ψ_{ABCD}. They satisfy equation (2.5.17), i.e.

$$\nabla^{AA'}\Psi_{ABCD} = 0. \qquad (4.1.6)$$

The common feature of all of these examples is that the principal part of the field equations involves a spinor divergence. We now have to show that the relevant field equations can be written in the form (4.1.2). For each of these, other than (4.1.3) there is a complication. Representations of zero rest mass fields which are irreducible under Lorentz transformations involve only symmetric spinors and so each of (4.1.4–6) represents an overdetermined or **constrained** system. For the moment we ignore this and concentrate on (4.1.4) ignoring the symmetry. Let us introduce a spin basis o, ι satisfying $o_A \iota^A = 1$. Then ϕ_{AB} can be replaced by the 4 (complex) scalars

$$\phi_0 = \phi_{AB}o^A o^B, \qquad \phi_1 = \phi_{AB}o^A \iota^B,$$
$$\phi_2 = \phi_{AB}\iota^A \iota^B, \qquad \phi_3 = \phi_{AB}\iota^A o^B. \qquad (4.1.7)$$

Projecting out the 4 equations (4.1.4) gives for the principal parts, in standard NP notation

$$\Delta\phi_0 - \delta\phi_3, \qquad \Delta\phi_1 - \delta\phi_2,$$
$$D\phi_2 - \bar{\delta}\phi_1, \qquad D\phi_3 - \bar{\delta}\phi_0. \tag{4.1.8}$$

Until now spinor indices have been written in abstract index notation and so have referred to no particular coordinate system. As will be shown in section 4.4 a coordinate system (u, v, x^α), where $\alpha = 2, 3$, can always be chosen so that the corresponding Newman-Penrose (NP) tetrad is of the form

$$l^a = o^A \bar{o}^{A'} = Q\partial/\partial v, \quad m^a = o^A \iota^{A'} = P^\alpha \partial/\partial x^\alpha,$$
$$n^a = \iota^A \iota^{A'} = \partial/\partial u + C^\alpha \partial/\partial x^\alpha. \tag{4.1.9}$$

Using the frame (4.1.8) we obtain a system with principal part

$$A^a_{\alpha\beta}\phi_{\beta,a}.$$

Here the coefficients are the matrices

$$A^u = \begin{pmatrix} 1 & 0 & 0 & 0 \\ 0 & 1 & 0 & 0 \\ 0 & 0 & 0 & 0 \\ 0 & 0 & 0 & 0 \end{pmatrix}, \qquad A^v = \begin{pmatrix} 0 & 0 & 0 & 0 \\ 0 & 0 & 0 & 0 \\ 0 & 0 & Q & 0 \\ 0 & 0 & 0 & Q \end{pmatrix},$$

$$A^A = \begin{pmatrix} C^A & 0 & 0 & -P^A \\ 0 & C^A & -P^A & 0 \\ 0 & -\bar{P}^A & 0 & 0 \\ -\bar{P}^A & 0 & 0 & 0 \end{pmatrix}. \tag{4.1.10}$$

Clearly $A^a_{\alpha\beta} = \bar{A}^a_{\beta\alpha}$ for each a, demonstrating that the system is symmetric. We have shown this result for a particular equation by an explicit calculation. However it is true for any equation whose principal part is a spinor divergence (Friedrich 1983, 1984). Although we chose a particular ordering of the equations and dependent variables, it is clear that the symmetry property is preserved under affine transformations of the dependent variables. Also, although our particular choice of tetrad diagonalizes A^u and A^v it is obvious that if the independent variables or the tetrad were changed the new $A^a_{\alpha\beta}$ would be linear combinations of the old ones, and symmetry would be maintained. To demonstrate hyperbolicity note that

$$H_{\alpha\beta} = A^u_{\alpha\beta} + Q A^v_{\alpha\beta}$$

is positive-definite. Symmetric hyperbolicity is independent of affine changes of dependent variables, and also of coordinate changes. Further if $\phi_{B...C}$ were a n-spinor, we would have obtained 2^n equations of precisely this type. The pattern is not restricted to our particular example.

This result is not directly applicable to Maxwell's theory. In that theory ϕ_{BC} is symmetric so that $\phi_3 = \phi_1$. We now have 4 equations (4.1.8) for 3 unknowns, i.e., the system is overdetermined or constrained. There are two different approaches. Friedrich, 1983, 1984, suggested setting $\phi_3 = \phi_1$ and considering the first, the sum of the second and fourth and the third displayed equation. The new coefficient matrices are

$$A^u = \begin{pmatrix} 1 & 0 & 0 \\ 0 & 1 & 0 \\ 0 & 0 & 1 \end{pmatrix}, \qquad A^v = \begin{pmatrix} 0 & 0 & 0 \\ 0 & Q & 0 \\ 0 & 0 & Q \end{pmatrix},$$

$$A^A = \begin{pmatrix} C^A & -P^A & 0 \\ -\overline{P}^A & C^A & -P^A \\ 0 & -\overline{P}^A & C^A \end{pmatrix},$$

clearly having the symmetry and hyperbolicity properties. However the difference of the second and fourth equations has not been considered. One may always choose initial data so that it is initially satisfied, but one must then verify that it is always satisfied. This is verified by a detailed calculation which will not be repeated here. An alternative approach, favoured by Stewart and Friedrich, 1982, is to set $\phi_3 = \phi_1$ and drop one equation, say the last. Deleting the last row and column of each matrix obviously does not change symmetry or hyperbolicity. Again it must be shown that the unused equation is satisfied identically as a consequence of the remaining equations provided that it is satisfied initially.

Our next example concerns a test perfect fluid. Both at the classical and relativistic levels this is a medium describable in terms of four variables, an *energy density* μ, a *pressure* $p(\mu)$ which is a given function of μ and either the three velocity components or a 4-velocity u^a satisfying $u^a u_a = 1$. The speed of sound will be denoted by c, where

$$c^2 = \frac{dp}{d\mu},$$

where, in the classical theory, the mass density is denoted by μ. We describe first the classical theory in two spatial dimensions, using coordinates (t, x, y). (The extension to three dimensions is then evident.) Let the velocity have components (u, v) and let the dependent variables be

$$\psi = (\mu, u, v)^T.$$

Then the well-known Euler equations can be written in the form (4.1.2) with

$$A^t = \begin{pmatrix} 1/\mu & 0 & 0 \\ 0 & \mu/c^2 & 0 \\ 0 & 0 & \mu/c^2 \end{pmatrix},$$

$$A^x = \begin{pmatrix} u/\mu & 1 & 0 \\ 1 & \mu u/c^2 & 0 \\ 0 & 0 & \mu u/c^2 \end{pmatrix},$$

$$A^y = \begin{pmatrix} v/\mu & 0 & 1 \\ 0 & \mu v/c^2 & 0 \\ 1 & 0 & \mu v/c^2 \end{pmatrix}.$$

The relativistic theory is just as straightforward. The energy momentum tensor is

$$T^{ab} = (\mu + p(\mu))u^a u^b - p(\mu)g^{ab}, \qquad (4.1.11)$$

and satisfies

$$\nabla_b T^{ab} = 0. \qquad (4.1.12)$$

We consider the case of a perfect fluid in a given spacetime governed by equations (4.1.11–12). We build in the constraint

$u^a u_a = 1$ by choosing as dependent variables $\psi_i = (\mu, u^\alpha)$ where in this paragraph $\alpha, \beta, \ldots$ range over $1, 2, 3$. It is convenient to introduce $v_\alpha = u_\alpha / u_0$. Then equations (4.1.12) are of the form (4.1.2) if we set

$$A^a_{00} = (\mu + p(\mu))^{-1} u^a, \qquad A^a_{\alpha 0} = A^a_{0\alpha} = \delta^a_\alpha - v_\alpha \delta^a_0,$$
$$A^a_{\alpha\beta} = (\mu + p(\mu)) c^{-2} \left(g_{00} v_\alpha v_\beta + g_{\alpha\beta} - 2 g_{0(\alpha} v_{\beta)} \right) u^a. \qquad (4.1.13)$$

(4.1.3) Problem. Verify equations (4.1.13).

So far all of these examples have involved "test fields", i.e., the influence of the field on the background has been ignored. The inclusion of back-reaction can, potentially, destroy symmetry. For Einstein's theory, with Ψ denoting the Weyl spinor and Φ the tracefree Ricci spinor, the Bianchi identities are, exercise 2.5.6

$$\nabla^A_{\ A'} \Psi_{ABCD} = \nabla^{B'}_{\ (B} \Phi_{CD)A'B'},$$
$$\nabla^{BB'} \Phi_{ABA'B'} = -3\nabla\Lambda. \qquad (4.1.14)$$

This is not of the form (4.1.4–6) and so Friedrich's theorem does not apply. Nevertheless symmetry and hyperbolicity hold for the conformal vacuum field equations, as will be explained in detail in section 4. This can be deduced heuristically as follows. In vacuum, equation (4.1.6) holds and so the left hand side of (4.1.14) would vanish. In a conformally related spacetime the Weyl spinor transforms homogeneously, and so the principal part of the left hand side of (4.1.14) vanishes. The vanishing of the right hand side is a (constrained) vanishing spinor divergence to which Friedrich's result can be applied.

It is not suggested that symmetry and hyperbolicity do not hold for more intimately coupled systems. However a-priori tests have not yet appeared, and until they do each case will need patient and detailed examination.

4.2   The Cauchy problem

In this section we shall for convenience drop the indices $\alpha, \beta, \ldots$ labelling the dependent variables. We remarked in section 4.1 that the system

$$A^a\psi_{,a} - B = 0, \qquad (4.2.1)$$

is symmetric hyperbolic if each matrix A^a is Hermitian and there exists a covector ξ_a such that $Q = A^a\xi_a$, is positive-definite.

(4.2.1) DEFINITION

*Suppose that for a given solution of (4.2.1), C is a hypersurface with normal ξ such that $Q = A^a\xi_a$ is positive-definite. Then C is said to be a **spacelike hypersurface for the system (and solution of) (4.2.1)**.*

There is a subtlety hidden in the nomenclature. If (4.2.1) is linear then whether or not C is spacelike depends only on the A^a and on C. However if (4.2.1) is merely quasilinear, the hyperbolicity of a given hypersurface depends also on the values of the dependent variables ψ in a neighbourhood of C.

Now suppose that a foliation of x-space M by spacelike hypersurfaces $t(x) = const.$ exists. We can choose $\xi_a = \nabla_a t$ and introduce new coordinates (t, y^i) where $i, j, k, \ldots = 2, \ldots, n$. In terms of the new coordinates we have

$$0 = A^a\psi_{,a} - B = Q\frac{\partial\psi}{\partial t} + (A^a y^i_{,a})\psi_{,i} - B,$$

where $Q = A^a\xi_a$ and $\psi_{,i} = \partial\psi/\partial y^i$. Now for quasilinear symmetric hyperbolic systems Q is symmetric and positive-definite. By means of an affine transformation in ψ-space we can convert Q to the unit matrix and the system can be written

$$\frac{\partial\psi}{\partial t} = C^i\psi_{,i} + D, \qquad (4.2.2)$$

where the C^i are Hermitian $N \times N$ matrices and D is a N-vector.

(4.2.2) DEFINITION

*Suppose that on (part of) a spacelike **initial** or **Cauchy surface** C, say $t = 0$ the unknowns ψ take on prescribed values $\psi_o(y)$.*

*The **Cauchy problem** is to determine in a suitable subset of the*
*future (i.e., the region $t > 0$) viz the **domain of dependence**,*
a solution $\psi(t, y)$ of (4.2.2) such that $\psi(0, y) = \psi_o(y)$.

Of course the existence and uniqueness of solutions of the
Cauchy problem depends on the smoothness of the coefficients and
data, and we will return to this below. Domains of dependence
are discussed in Courant and Hilbert, 1962, chapter 6, section 7.

Consider first a linear system so that C^i is independent of ψ
and D is linear in ψ. A simple "energy argument" can be used to
demonstrate uniqueness of solutions of the (linear) Cauchy problem. (A detailed account can be found in Courant and Hilbert,
1962, chapter 6.) The simplest problem is when the coefficients in
(4.2.2) are analytic functions of t, y^i and the data ψ_o are poly-
nomials in the y^i. On C we know ψ and (4.2.2) shows that
$\partial\psi/\partial t = C^i \psi_{o,i} + D$ there. By repeated differentiation of the
data with respect to the y^i one can build up $\partial^n \psi/\partial t^n$ on C. Now
one can construct a formal Taylor series with respect to t for ψ.
This can be shown to converge uniformly with respect to y^i in a
strip $t \in [0, t_o]$ in x-space M. This is the **Cauchy-Kovalevskaya**
theorem, Courant and Hilbert, 1962, chapter I, section 7.

The restriction to analytic coefficients and data is of course
hopelessly unrealistic. We have to introduce the concept of
Sobolev spaces. (See Hawking and Ellis, 1973, section 7.4 for
a succinct introduction.) Let $\hat{g}$ be a positive-definite metric on M
and let $\mathcal{N}$ be an imbedded submanifold with compact closure in
M, and let $d\sigma$ be the volume element of $\mathcal{N}$ induced by $\hat{g}$. Let $D^p \psi$
denote the p'th derivatives of ψ. We define the **Sobolev norms**
via

$$\|\psi, \mathcal{N}\|_m = \left[\sum_{p=0}^{m} \int_{\mathcal{N}} |D^p \psi|^2 \, d\sigma \right]^{1/2} . \qquad (4.2.3)$$

If the derivatives are restricted to those intrinsic to $\mathcal{N}$ the corresponding expression is denoted by a tilde on the final norm symbol.
The **Sobolev spaces** $\mathcal{W}^m(\mathcal{N})$ are then the completions of the
space of C^∞ functions with respect to the Sobolev norms. The
relation to the more usual maximum norm is as follows. Suppose

$m > \dim(\mathcal{N})/2$. Then

$$\|\psi\| < const. \times \|\psi, \mathcal{N}\tilde{\|}_m. \qquad (4.2.4)$$

In particular if $m > \dim(\mathcal{N})/2 + s + 1$, elements of $\mathcal{W}^m(\mathcal{N})$ are fields on $\mathcal{N}$ with bounded partial derivatives of order up to s. Thus $\mathcal{W}^m(\mathcal{N})$ is a Banach space in which the C^∞ fields form dense subsets. In other words for large m, $\mathcal{W}^m(\mathcal{N})$ is the set of smooth fields on $\mathcal{N}$.

Now the analytic functions in $\mathcal{N}$ are dense in $\mathcal{W}^m(\mathcal{N})$. Thus any element $\psi_o \in \mathcal{W}^m(\mathcal{C})$ can be approximated by a sequence ψ_{no} of analytic data. By an energy argument the analytic solutions ψ_n of the corresponding Cauchy problems converge to an element $\psi \in \mathcal{W}^m(\mathcal{M})$. This may only exist locally, and so we would want to be able to choose a new Cauchy surface and start again, thus constructing a global solution by patchwork. For this to be possible we need the data and coefficients to be in $\mathcal{W}^m(\mathcal{C})$, $\mathcal{W}^m(\mathcal{M})$ respectively with $m > \dim(M)/2 + 1$. (The reason for this is ultimately (4.2.4). See Courant and Hilbert, 1962, chapter 6, section 10 for a further general discussion, and the two references cited below for a more technically detailed treatment.)

Quasilinear systems have been treated in detail by a number of authors. Two discussions which are relevant here are those by Fischer and Marsden, 1972, and by Kato, 1975. The basic idea is to utilize the theory of linear systems as follows. The quasilinear system

$$\frac{\partial \psi}{\partial t} = C^i(\psi)\psi_{,i} + D(\psi),$$

is replaced by a sequence of linear problems

$$\frac{\partial \psi_{n+1}}{\partial t} = C^i(\psi_n)\psi_{n+1,i} + E(\psi_n)\psi_{n+1} + F(\psi_n), \quad n \geq 0, \quad (4.2.5)$$

all with data ψ_o. In other words we solve a succession of linear problems in which the coefficients are determined by the solution of the previous problem. We denote this operation by

$$\psi_{n+1} = T(\psi_n). \qquad (4.2.6)$$

If T has a fixed point ψ, i.e., $\psi = T(\psi)$, then this is clearly a solution of the (quasilinear) Cauchy problem. (Since we are dealing with complete spaces it is sufficient to show that T is a contraction mapping.) Of course one then has to show that the sequence ψ_n is well-defined and in particular that ψ_{n+1} has the same smoothness as ψ_n. Thus we need the data to be in $\mathcal{W}^m(\mathcal{C})$ and the coefficients to be in $\mathcal{W}^m(\mathcal{M})$ where $m > \dim(\mathcal{M})/2 + 1$. Provided that these smoothness conditions are satisfied, uniqueness, stability and local existence of solutions of the Cauchy problem is guaranteed.

4.3   The characteristic initial value problem

Let $\phi(x)$ be a scalar field on M, and let $p_a = \nabla_a \phi$ be its gradient. With each first order quasilinear equation

$$A^a_{\alpha\beta}\psi_{\beta,a} = B_\alpha, \qquad (4.3.1)$$

we may associate a $N \times N$ matrix

$$Q_{\alpha\beta} = A^a_{\alpha\beta}p_a. \qquad (4.3.2)$$

In section 4.1 we defined (4.3.1) to be hyperbolic if surfaces $\phi = const.$ existed for which (4.3.2) was positive-definite. Of course other possibilities can exist. If $Q_{\alpha\beta}$ is non-singular and not positive definite the surface is *timelike for equation (4.3.1)*. If $Q_{\alpha\beta}$ is singular,

$$Q \equiv \det(Q_{\alpha\beta}) = 0, \qquad (4.3.3)$$

then $\phi = const.$ is a *characteristic hypersurface* or *null hypersurface for equation (4.3.1)*. Again, just as in section 4.2 we note that if (4.3.1) is quasilinear rather than linear, whether or not a given hypersurface is timelike or characteristic depends not only on the equation (4.3.1) and the hypersurface, but also on the values taken by the dependent variables ψ_β on the surface.

Q can be regarded as a homogeneous polynomial of degree N in the p_a with coefficients which depend on x. For fixed x we may solve the characteristic condition $Q = 0$ to obtain a family of normals or **null directions** at x, which define the **Monge cone** at x.

The following construction will prove particularly fruitful. Let C be a spacelike hypersurface for (4.3.1) and let S be a hypersurface within C, so that S has codimension 2. Now at each point $x \in S$ we ask for the set of null covectors at x which are orthogonal to all vectors tangent to S, i.e., the set of p_a such that

$$Q(x, p) = 0, \qquad p_a v^a = 0 \quad \text{for all } v^a \in T_x(S). \qquad (4.3.4)$$

Because (4.3.4) contains $n - 1$ conditions and they are homogeneous in the p_a, solutions p_{oa} will consist of a discrete set of null directions at each $x_o \in S$.

Next consider $Q(x, p)$ as a Hamiltonian for a totally fictitious dynamical system. Hamilton's equations are

$$\frac{dx^a}{ds} = \frac{\partial Q}{\partial p_a}, \qquad \frac{dp_a}{ds} = -\frac{\partial Q}{\partial x^a}, \qquad (4.3.5)$$

where s is fictitious time. As initial data we take

$$x^a(0) = x_o^a, \qquad p_a(0) = p_{oa}, \qquad (4.3.6)$$

where $x_o \in S$ and p_{oa} is a solution of (4.3.4) which is continuous with respect to x_o. At least locally (for the ordinary differential equations are nonlinear) (4.3.5–6) have a solution, and as the initial point x_o varies on S this defines a hypersurface $\mathcal{N}$ containing S. Note further that by construction $Q = 0$ at those points of $\mathcal{N}$ which lie on S. Now the Hamiltonian Q is time-independent and so is a constant of the motion. Thus $Q \equiv 0$ on $\mathcal{N}$, i.e., $\mathcal{N}$ is a characteristic surface. The integral curves of (4.3.5) or generators of $\mathcal{N}$ are called **bicharacteristic curves**. (Many textbooks refer to them as characteristic curves; this is perhaps a legacy of the oversimplification in which only two independent variables occur, so that characteristic hypersurfaces and bicharacteristic curves coincide.) Finally we evaluate the change in ϕ along the bicharacteristics. Clearly

$$\frac{d\phi}{ds} = p_a \frac{dx^a}{ds} = p_a \frac{\partial Q}{\partial p_a}.$$

Now Q is a homogeneous polynomial in the p_a. By Euler's theorem the last term above is proportional to Q and so vanishes. Thus ϕ is constant along the generators of $\mathcal{N}$. In particular if ϕ is constant on S then it is constant on $\mathcal{N}$.

In section 4.2 we saw that if data ψ_β was given on a spacelike hypersurface $C : t = const.$, our general quasilinear system (4.2.2) specified the the normal derivatives, leading to the Cauchy problem. For characteristic surfaces the situation is totally different. Suppose we construct a local family of characteristic surfaces $u = const.$ perhaps by the above procedure. We next introduce new coordinates (u, y^i), where $i, j, k, \ldots = 2, \ldots, n$. In terms of the new coordinates we have, using (4.3.2)

$$0 = A^a_{\alpha\beta}\psi_{\beta,a} - B_\alpha = Q_{\alpha\beta}\psi_{\beta,u} + (A^a_{\alpha\beta}y^i_{,a})\psi_{\beta,i} - B_\alpha. \qquad (4.3.7)$$

By construction $Q_{\alpha\beta}$ is both symmetric and singular. Thus it possesses $m_u > 0$ null eigenvectors $\ell_\alpha(x)$, i.e., $Q_{\alpha\beta}\ell_\alpha = 0$. We next multiply (4.3.7) by such a ℓ_α finding

$$(\ell_\alpha A^a_{\alpha\beta}y^i_{,a})\psi_{\beta,i} = \ell_\alpha B_\alpha. \qquad (4.3.8)$$

This equation contains no u-derivatives and so is ***intrinsic*** to the surface $u = const.$ Thus the system (4.3.1) contains m_u equations intrinsic to each $u = const.$ characteristic surface. In particular data cannot be given freely on an initial surface, say $u = 0$. At most one may specify $N - m_u$ pieces of data freely. In all the cases of physical interest we can make an internal transformation of the dependent variables to arrange that these correspond to $N - m_u$ dependent variables, and we call these the u-***data*** D_u. The remaining m_u dependent variables will be called u-***variables*** V_u. They can be specified freely only on a hypersurface within $u = 0$ for there are equations governing their evolution within $u = 0$. Thus the situation is considerably more complicated than that for a spacelike initial surface.

We can now develop the idea of a characteristic initial value problem. Let C be a spacelike hypersurface for equation (4.3.1) (not forgetting the subtlety when (4.3.1) is quasilinear) and let S_0 be a codimension 2 submanifold in C which divides C locally into two parts which we call 'left' and 'right'. Consider first those 'right-directed' characteristic surfaces through S_0, i.e., those for which the projection of the bicharacteristic direction into $\mathcal{C}$ points to the 'right'. One of these will have the property that all of the others lie to its future. It will be called $\mathcal{N}_0$. Next consider a 1-parameter foliation of $\mathcal{C}$ by codimension 2 submanifolds which include S_0. By repeating the construction we obtain a 1-parameter family of characteristic surfaces $\mathcal{N}_u$. A scalar field u can then be defined via the requirement $u = const.$ on $\mathcal{N}_u$. A similar construction can be performed on the 'left-directed' characteristic hypersurfaces, leading to another foliation $\mathcal{N}'_v$ and a scalar field v. We next introduce coordinates (u, v, y^i) with $i, j, k, \ldots = 3, \ldots, n$ and the y^i as arbitrary coordinates on each codimension 2 submanifold $S_{uv} : u = const, v = const$. Finally we determine the sets D_u, D_v of dependent variables which comprise the u-data and v-data respectively. Provided the complement of $V_u \cup V_v$ is empty we may state a characteristic initial value problem as follows. Suppose the u-data D_u is specified freely on $\mathcal{N}_0$, the v-data D_v is specified freely on $\mathcal{N}'_0$ and the complement of their union is specified on $S_0 = \mathcal{N}_0 \cap \mathcal{N}'_0$; determine the solution in $\mathcal{M}$, the future of $\mathcal{N}_0 \cup \mathcal{N}'_0$.

We illustrate these concepts by considering a simple example, system (4.1.7–8). It is clear that on $\mathcal{N}_0$, $u = 0$, ϕ_0, ϕ_1 can be specified freely and so form the u-data D_u. Similarly ϕ_2, ϕ_3 can be specified freely on $\mathcal{N}'_0$ and so form the v-data D_v. Thus the u-variables are $V_u = \{\phi_2, \phi_3\}$, the v- variables are $V_v = \{\phi_0, \phi_1\}$ and $D_u \cap D_v = \emptyset$. However if we convert (4.1.8) to Maxwell's equations by setting $\phi_3 = \phi_1$ a priori then $V_u = \{\phi_1, \phi_2\}$ as before, but $D_u = \{\phi_0\}$. The v-variables are $V_v = \{\phi_0, \phi_1\}$ and so $D_v = \{\phi_2\}$. Note that the complement of their union is no longer empty. (This is typical of constrained systems.) Data for ϕ_1 can be given freely neither on $\mathcal{N}_0$ nor $\mathcal{N}'_0$, but only on S_0.

Having stated an initial value problem we now examine its resolution. We shall assume that x-space has been foliated by two families of characteristic surfaces $\mathcal{N}_u : u = const.$ and $\mathcal{N}'_v : v = const.$ chosen as above and that the coordinate system (u, v, y^i) is being used. Further u-data D_u is given on $\mathcal{N}_0$ and

v-data D_v is given on $\mathcal{N}_0'$. Finally if the complement of $D_u \cup D_v$ is not empty this remaining data is given on $S_0 = \mathcal{N}_0 \cap \mathcal{N}_0'$. We shall demonstrate informally the existence and uniqueness of solutions of the system (4.3.1) in $\mathcal{M}$ the future of $\mathcal{N}_0 \cup \mathcal{N}_0'$. In order to make the discussion more concrete we shall continue to develop the example started above.

Consider first $\mathcal{N}_0 : u = 0$. Here we are specifying freely the u-data D_u. Since the v-data are given freely on $\mathcal{N}_0'$ they are known on S_0. Further the complement of $D_u \cup D_v$ is given on S_0 and so the v-variables are known on S_0. (In our example ϕ_0, ϕ_1 are u-data and v-variables while ϕ_2, ϕ_3 are v-data and u-variables. In the constrained version $V_u = \{\phi_1, \phi_2\}$, $V_v = \{\phi_0, \phi_1\}$, ϕ_0 is the u-data, ϕ_2 is the v-data and ϕ_1 is specified on S_0. For convenience we may delete ϕ_1 from the set V_v.)

Now consider a hypersurface $\phi = const.$ which is initially the 'right' half of the Cauchy surface $\mathcal{C}$ containing S_0, and which evolves to become $\mathcal{N}_0$. Initially $Q_{\alpha\beta} = A^a_{\alpha\beta}\phi_{,a}$ is nonsingular, and by construction first becomes singular when $\mathcal{N}_0$ is reached. At this surface m_u of the initially positive eigenvalues of $Q_{\alpha\beta}$ will have been reduced to zero. Without loss of generality we may work with a set of dependent variables for which $Q_{\alpha\beta}$ is diagonal, and the last m_u eigenvalues are zero. We now strike out the first $N - m_u$ equations from the system. What remains is the reduced system (4.3.7). This has the property that the coefficient matrix multiplying the derivatives has maximal rank m_u, and if we regard the u-data as known, consists of m_u equations for m_u unknowns, the v-variables. By the construction given above, data for these quantities is known on S_0 which is clearly spacelike for the reduced system.

Thus we have a standard Cauchy problem with $n - 1$ independent variables v, y^i and m_u dependent ones. By the theory of section 4.2 a solution can be constructed locally. (In our example $Q_{\alpha\beta}$ is the matrix A^u in equation (4.1.10). This is already diagonal and clearly $m_u = 2$. If we strike out the first two equations and regard ϕ_0, ϕ_1 as given then the coefficient matrix is the lower right 2×2 submatrix. This is obviously positive-definite.) An almost identical construction can be carried out on $\mathcal{N}_0'$ by interchanging the u- and v- quantities so that the solution is known on the complete initial surface $\mathcal{N}_0 \cup \mathcal{N}_0'$. We now proceed to determine the solution to the future of this initial surface.

Now suppose the solution is known on $\mathcal{N}_u \cup \mathcal{N}_v'$. In particular the solution is known on $S_{u+du,v}$ and on $S_{u,v+dv}$. We shall demonstrate how to construct the solution on $S_{u+du,v+dv} = \mathcal{N}_{u+du} \cap \mathcal{N}_{v+dv}'$, where du, dv are small. Once this has been done repetition of the process produces the solution on $\mathcal{N}_{u+du}$. By a similar operation we can construct the solution on $\mathcal{N}_{v+dv}'$, and thus on $\mathcal{N}_{u+du} \cup \mathcal{N}_{v+dv}'$. Successive patchwork then builds a solution in $\mathcal{M}$.

By assumption we know the solutions on $S_{u,v+dv}$ and on $S_{u+du,v}$. $\mathcal{N}_{u+du}$ is a characteristic surface and within it we have a reduced set of m_u equations for v-variables. By construction this set is hyperbolic. Given *analytic* data on $S_{u+du,v}$ we can construct the solution of this reduced set on $S_{u+du,v+dv}$. This is (of course) precisely the Cauchy-Kovalevskaya theory for the standard Cauchy problem. There is a similar construction for the u-variables on $\mathcal{N}_{v+dv}'$. Note that some adjustments to the values of du and dv may be necessary to get both sets of Taylor series to converge, but they will for sufficiently small values. Thus the Cauchy-Kovalevskaya construction works for the characteristic initial value problem. It is often alleged that this was first proved by Duff, 1958, and indeed the ideas used here are based on that work. However he considered linear systems with data specified on an initial surface part of which was spacelike and part characteristic. An explicit demonstration of the construction has been given by Friedrich, 1982, for the asymptotic characteristic initial value problem for the conformal vacuum Einstein field equations.

Next consider the case of finite differentiability for the data, coefficients and solution. First we consider the linear case. Using the data we construct a consistent approximation for the u-variables on $\mathcal{N}_{u+du}$ and construct the solution to the Cauchy problem for the v-variables on $[v, v+dv]$. Using the data and this solution we construct a consistent approximation for the v-variables on $\mathcal{N}_{v+dv}'$ and solve the Cauchy problem for the u-variables on $[u, u+du]$. Because of the established theory for the Cauchy problem such a process is guaranteed to produce a new set of variables of the appropriate differentiability class at $\mathcal{S}_{u+du,v+dv}$. We now iterate this process in a manner similar to that used in the transition from linear to quasilinear systems in section 4.3. It is extremely plausible that for small enough du, dv the process is a contraction mapping with a fixed point, and it should not be too difficult to write out a formal proof. We assume that this can be done. To progress to the quasilinear case we merely combine the iteration above with that used in section 4.2 for the Cauchy problem.

Of course other approaches to the characteristic initial value are possible. An interesting reduction to the Cauchy problem has been given by Rendall, 1990, who cites earlier approaches.

4.4   The conformal vacuum field equations of general relativity

In this section we show how the conformal vacuum field equations of general relativity can be written as a symmetric first order system. The physically interesting application of the Einstein vacuum field equations is to the field outside an isolated gravitating body such as a cluster of stars. There is a well-defined theory of the asymptotics of isolated systems, which has been discussed in chapter 3. Let the true or *physical vacuum* spacetime have metric $\tilde{g}_{\mu\nu}$. (In this section we shall assume that spacetime M has a (local) coordinate chart (x^{μ}) with greek indices ranging over 0–3. This is not the convention that is used throughout the rest of the book, but it will be used here to improve the legibility of the subsequent treatment.) The key idea is to use a *conformal transformation* introduced in chapter 3

$$\tilde{g}_{\mu\nu} \mapsto g_{\mu\nu} = \Omega^2 \tilde{g}_{\mu\nu}, \qquad (4.4.1)$$

where Ω is a non-negative scalar field on M. By letting $\Omega \to 0$ as "infinity" is approached one can scale distances so that "infinity" is at a finite distance as measured by the rescaled metric. The 4-manifold whose metric is $g_{\mu\nu}$ is called the *unphysical spacetime* and $g_{\mu\nu}$ is the *unphysical metric*. Of course a transformation (4.4.1) will change the curvature. The Weyl tensor components transform homogeneously, i.e., they become multiplied by appropriate powers of Ω, see e.g., equation (3.4.5). However the Ricci tensor components transform in an inhomogeneous and complicated manner, given by (3.4.8–9). Thus even though the Ricci tensor vanishes for the physical vacuum spacetime, it will not do so in the conformally rescaled spacetime. The latter contains "unphysical matter", and this greatly complicates the discussion. However the reader who is uninterested in asymptopia can recover the standard Einstein vacuum theory by setting $\Omega = 1$, and deleting all derivatives of Ω and Ricci spinor components from the following discussion.

The theory presented here appeared first in Friedrich, 1979, 1981a, 1981b, 1982, 1983, 1984, and his approach will be followed closely. As usual capital roman letters $A, B, \ldots$ will denote abstract spinor indices. We shall also need a local spinor dyad, and dyad indices will be denoted by lower case roman letters $a, b, \ldots$. This is considerably easier to read than some conventions, e.g., Penrose and Rindler, 1984, 1986, but it does mean relabelling the spacetime coordinate chart. Fortunately spacetime coordinate indices will appear rarely.

Let ϵ_a^A denote a local spinor dyad such that

$$\epsilon_{AB}\epsilon_a^A\epsilon_b^B = \epsilon_{ab},$$

where $\epsilon_{AB}, \epsilon_{ab}$ denote the usual symplectic 2-form. (For example we may choose $\epsilon_0^A = o^A$, $\epsilon_1^A = \iota^A$.) The dual dyad will be denoted ϵ_A^a. Let $e_{aa'}$ denote the corresponding Newman-Penrose (NP) tetrad. Its components with respect to the coordinate-induced basis of the tangent space are $E_{aa'}^\mu = e_{aa'}(x^\mu)$, or equivalently

$$e_{aa'} = E_{aa'}^\mu \partial/\partial x^\mu. \tag{4.4.2}$$

The $E_{aa'}^\mu$ are usually called $l^\mu, n^\mu, m^\mu, \overline{m}^\mu$ and define the metric tensor components via

$$g^{\mu\nu} = 2l^{(\mu}n^{\nu)} - 2m^{(\mu}\overline{m}^{\nu)}. \tag{4.4.3}$$

They are our fundamental dependent variables. (Problems of gauge freedom will be addressed shortly.)

Let ϕ be any scalar field. The **torsion tensor** T was defined in section 1.9 and can be expressed as

$$2\nabla_{[\alpha}\nabla_{\beta]}\phi = T_{\alpha\beta}{}^{\gamma}\nabla_{\gamma}\phi. \qquad (4.4.4)$$

If we introduce the spinor

$$t_{AB}{}^{CC'} = \tfrac{1}{2}T_{AB}{}^{C}{}_{A'}{}^{A'C'},$$

(4.4.4) becomes

$$\Box_{AB}\phi \equiv \nabla_{A'(A}\nabla_{B)}{}^{A'}\phi = t_{AB}{}^{CC'}\nabla_{CC'}\phi, \qquad (4.4.5)$$

or

$$\nabla_{AA'}\nabla_{B}{}^{A'}\phi \equiv t_{AB}{}^{CC'}\nabla_{CC'}\phi + \tfrac{1}{2}\epsilon_{AB}\Box\,\phi, \qquad \forall\phi. \qquad (4.4.6)$$

Henceforth we shall assume zero torsion. We now want to apply the identity (4.4.6) to the scalar fields x^{μ}. We first define

$$g^{\mu} = \tfrac{1}{2}\Box\,x^{\mu}. \qquad (4.4.7)$$

Now for any practical purpose one needs dyad rather than abstract index equations. Adopting the standard Penrose and Rindler, 1984, definition of the **spin-coefficients**

$$\gamma_{aa'c}{}^{b} \equiv \epsilon_{A}^{b}\nabla_{aa'}\epsilon_{c}^{A} = -\epsilon_{c}^{A}\nabla_{aa'}\epsilon_{A}^{b},$$

equation (4.4.6) becomes

$$\nabla_{aa'}E^{\mu}{}_{b}{}^{a'} = \gamma_{aa'b}{}^{c}E^{\mu}{}_{c}{}^{a'} + \overline{\gamma}_{aa'}{}^{a'c'}E^{\mu}{}_{bc'} + \epsilon_{ab}g^{\mu}. \qquad (4.4.8)$$

The equations (4.4.8) form a symmetric quasilinear first order system governing the propagation of our fundamental variables, the $E^{\mu}{}_{aa'}$, provided that the spin-coefficients are known. Note the rôle of the g^{μ} here. Had we transcribed (4.4.5) into dyad form directly we would have obtained an equation containing a spinor divergence with a further symmetrization applied. (Note in particular the brackets surrounding the indices in the middle term of (4.4.5).) The theorem of Friedrich, 1983, 1984, does not apply to such systems. The extra term in the identity (4.4.6) was added to remove this symmetrization. There are no equations governing the g^μ defined by (4.4.7). They can therefore be regarded as four arbitrary functions to be chosen freely. In conventional accounts they are usually referred to as *coordinate gauge freedom.* A judicious choice can often effect remarkable simplifications in the dyad equations.

Next we need equations governing the propagation of the spin-coefficients. The spinor Ricci identities, equation (4.11.5) of Penrose and Rindler, 1984, furnish equations for the symmetrized derivatives $\nabla_{a'(a}\gamma_{b)}{}^{a'}{}_{cd}$ and $\nabla_{a(a'}\gamma^a{}_{b')cd}$. To remove the symmetrizations we introduce the symmetric spinor field

$$f_{cd} = f_{(cd)} = \tfrac{1}{2}\nabla_{aa'}\gamma^{aa'}{}_{cd}. \qquad (4.4.9)$$

The Ricci identities then become

$$
\begin{aligned}
\nabla_{aa'}\gamma_b{}^{a'}{}_{cd} &= \gamma_{a'(a|c|}{}^e\,\gamma_{b)}{}^{a'}{}_{ed} + \gamma_{a'(ab)}{}^e\,\gamma_e{}^{a'}{}_{cd} + \\
&\quad \overline{\gamma}_{a'(a}{}^{a'e'}\gamma_{b)e'cd} + \Psi_{abcd} - \\
&\quad 2\Lambda\epsilon_{a(c}\epsilon_{d)b} + \epsilon_{ab}f_{cd},
\end{aligned}
\qquad (4.4.10)
$$

$$
\begin{aligned}
\nabla_{aa'}\gamma^a{}_{b'cd} &= \gamma_{a(a'|c|}{}^e\,\gamma^a{}_{b')ed} + \gamma_{a(a'}{}^{ae}\,\gamma_{b')ecd} + \\
&\quad \overline{\gamma}_{a(a'b')}{}^{e'}\gamma^a{}_{e'cd} + \Phi_{cda'b'} + \epsilon_{a'b'}f_{cd},
\end{aligned}
$$

where Ψ_{abcd}, $\Phi_{aba'b'}$ and Λ are the dyad components of the Weyl spinor, Ricci spinor and Ricci scalar. The rôle of f_{ab} here is precisely the same as that of g^μ in equation (4.4.8), and there is a similar interpretation. We have not made a particular choice of frame. The freedom available is of course the 6-parameter Lorentz group. Note that f_{ab} is symmetric and so contains 3 complex functions of position, equivalent to the frame freedom. In other words f_{ab} describes the *tetrad gauge freedom.* Fortunately there is no need to compute explicit forms for the equations (4.4.10). A complete set for the case $f_{ab} = 0$ is given in appendix B. It is very easy to add in the f_{ab} terms. Because the principal parts of equations (4.4.10) are spinor divergences we have a symmetric first order quasilinear system with which to propagate the spin-coefficients. However the right hand sides involve the curvature spinors and the curvature scalar. We must therefore construct equations for their evolution.

Consistent with equation (3.4.2) we set

$$\epsilon_{AB} = \Omega \tilde{\epsilon}_{AB}. \tag{4.4.11}$$

We denote the first derivatives of Ω by

$$s_{AA'} = \nabla_{AA'}\Omega. \tag{4.4.12}$$

It is then straightforward to transcribe equations (3.4.5–7) as

$$\begin{aligned}
\tilde{\Psi}_{ABCD} &= \Psi_{ABCD}, \\
\tilde{\Lambda} &= \Omega^2 \Lambda - \tfrac{1}{4}\Omega\nabla_{AA'}s^{AA'} + \tfrac{1}{2}s_{AA'}s^{AA'}, \\
\tilde{\Phi}_{ABA'B'} &= \Phi_{ABA'B'} + \Omega^{-1}\nabla_{A(A'}s_{B')B}.
\end{aligned} \tag{4.4.13}$$

First we impose the Bianchi identity in the physical vacuum. This can be written as

$$\tilde{\nabla}^A{}_{A'}\tilde{\Psi}_{ABCD} = 0,$$

or

or

$$\nabla^A{}_{A'}(\Omega^{-1}\Psi_{ABCD}) = 0.$$

It is convenient to define

$$\phi_{ABCD} = \Omega^{-1}\Psi_{ABCD}. \tag{4.4.14}$$

Although Ω vanishes at null infinity so does Ψ_{ABCD} and ϕ_{ABCD} is regular there. Thus the physical Bianchi identities become

$$\nabla^A{}_{A'}\phi_{ABCD} = 0,$$

or

$$\nabla^a{}_{a'}\phi_{abcd} = \gamma^a{}_{a'}{}^e{}_a\phi_{ebcd} + 3\gamma^a{}_{a'}{}^e{}_{(b}\phi_{cd)ae}. \tag{4.4.15}$$

This is a spinor divergence and so generates a symmetric quasilinear system for the (rescaled) Weyl spinor components. These equations can be written out explicitly by taking the Bianchi identities in appendix B and deleting the Ricci terms Φ and Λ.

Next we impose the Bianchi identity in the unphysical space-time. This reads

$$\nabla_A{}^{A'}\Phi_{CDA'B'} = \nabla^B{}_{B'}\Psi_{ABCD} + \epsilon_{A(C}\nabla_{D)B'}\Lambda$$
$$= \phi_{ABCD}s^B{}_{B'} + \epsilon_{A(C}\nabla_{D)B'}\Lambda, \quad (4.4.16)$$

where the physical Bianchi identity has been used. In dyad components we have

$$\nabla_a{}^{a'}\Phi_{cda'b'} = 2\gamma_a{}^{a'e}{}_{(c}\Phi_{d)ea'b'} + 2\overline{\gamma}_a{}^{a'e'}{}_{(a'}\Phi_{b')e'cd} + $$
$$\phi_{abcd}s^b{}_{b'} + 2\epsilon_{a(c}\nabla_{d)b'}\Lambda. \quad (4.4.17)$$

This is a spinor divergence and hence a symmetric quasilinear system for the Ricci spinor components, but we need in addition propagation equations for $s_{aa'}$ and Λ.

To obtain an equation for $s_{aa'}$ we impose part of the physical vacuum Einstein equation, viz.

$$\tilde{\Phi}_{ABA'B'} = 0, \quad (4.4.18)$$

or, using (4.4.13),

$$\nabla_{A(A'}s_{B')B} = -\Omega\Phi_{ABA'B'}. \quad (4.4.19)$$

This is not quite in its final form. Because there is no torsion $\nabla_{A(A'}s_{B')}{}^A = 0$, and introducing

$$C = \tfrac{1}{4}\Box\,\Omega \equiv \tfrac{1}{4}\nabla_{AA'}s^{AA'}, \quad (4.4.20)$$

we have

$$\nabla_{aa'}s_{bb'} = \gamma_{aa'}{}^c{}_b s_{cb'} + \overline{\gamma}_{aa'}{}^{c'}{}_{b'}s_{bc'} - \Omega\Phi_{aba'b'} + C\epsilon_{ab}\epsilon_{a'b'}. \quad (4.4.21)$$

Thus each derivative of $s_{aa'}$ is computable. This is not symmetric although a certain subset forms a symmetric first order quasilinear system for $s_{aa'}$. A new quantity C has been introduced. However its derivative is

$$4\nabla_{AA'}C = \nabla_{AA'}\nabla_{BB'}s^{BB'}$$
$$= \nabla_{BB'}\nabla_{AA'}s^{BB'} + \epsilon_{A'B'}\Box_{AB}s^{BB'} + \epsilon_{AB}\Box_{A'B'}s^{BB'}$$
$$= \nabla_{BB'}\left[-\Omega\Phi_A{}^B{}_{A'}{}^{B'} + C\epsilon_A{}^B\epsilon_{A'}{}^{B'}\right] + $$
$$\epsilon_{A'B'}\Box_{AB}s^{BB'} + \epsilon_{AB}\Box_{A'B'}s^{BB'}$$
$$= 3\Omega\nabla_{AA'}\Lambda - 3\Phi_{ABA'B'}s^{BB'} + 6\Lambda s_{AA'} + \nabla_{AA'}C,$$

after some work. Thus

$$\nabla_{AA'}C = \Omega\nabla_{AA'}\Lambda - \Phi_{ABA'B'}s^{BB'} + 2\Lambda s_{AA'}, \qquad (4.4.22)$$

and

$$\nabla_{aa'}C = \Omega\nabla_{aa'}\Lambda - \Phi_{aba'b'}s^{bb'} + 2\Lambda s_{aa'}. \qquad (4.4.23)$$

We still need an equation governing the propagation of Λ. The remainder of the physical vacuum Einstein equation is $\tilde{\Lambda} = 0$, or from (4.4.13)

$$2\tilde{\Lambda} \equiv 2\Omega^2\Lambda - 2\Omega C + s_{aa'}s^{aa'} = 0. \qquad (4.4.24)$$

Note that the physical vacuum (contracted) Bianchi identities imply $\nabla_\nu\tilde{R} = 0$ or $\nabla_{aa'}\tilde{\Lambda} = 0$. Thus equation (4.4.24) only needs to be imposed initially. It should be noted that there is some freedom. Clearly we may replace Ω in (4.4.1) by $\theta\Omega$ where θ is any strictly positive scalar field. This *conformal gauge freedom* can be used to specify Λ. We shall usually use it to set $\Lambda = 0$ but other choices are possible.

This has been a long argument and so we summarize it as follows. We have set up an elaborate symmetric quasilinear first order system of equations for our dependent variables according to the following scheme.

variables	equations
$E^\mu{}_{aa'}$	(4.4.8)
$\gamma_{aa'bc}$	(4.4.10)
ϕ_{abcd}	(4.4.15)
$\Phi_{aba'b'}$	(4.4.17)
Ω	(4.4.12)
$s_{aa'}$	(4.4.21)
C	(4.4.23)

In addition (4.4.24) is an initial constraint which only has to be imposed at one point. The gauge freedom available in this scheme is as follows.

gauge	free choice
coordinate	g^μ
tetrad	f_{ab}
conformal	Λ

We have established that our scheme of equations forms a symmetric quasilinear first order system. We need also to establish hyperbolicity, and in order to do this we have to choose a particular coordinate system and tetrad. Of course the hyperbolicity or otherwise of a system is a coordinate- and tetrad-independent property, and so this is no real restriction. Our first two coordinates are the characteristic coordinates u, v of section 4.3. On the initial 2-surface $S = \mathcal{N}_0 \cap \mathcal{N}_0'$ we choose arbitrary coordinates x^A, where for the moment $A, B, \ldots = 2, 3$. We propagate these coordinates onto $\mathcal{N}_0'$ by requiring them to be constant along the generators. Once the x^A are defined on $\mathcal{N}_0'$ we propagate them into $\mathcal{M}$ by requiring them to be constant along the generators of each $\mathcal{N}_u$. This clearly establishes a coordinate system in $\mathcal{M}$. We choose vector fields l, n to be tangent to the generators of $\mathcal{N}_v, \mathcal{N}_u'$ respectively, which makes both of them null, and to be normalized via $l.n = 1$. There is still some **boost freedom**, $l \mapsto \lambda l,\ n \to \lambda^{-1} n$ with λ real. We use this up to set $l_\mu = \nabla_\mu u$. There is still some freedom left, for we may choose a relabelling $u \mapsto U(u)$. It is advantageous to reserve this freedom for ameliorating particular problems. We can always choose the vector fields $m, \overline{m}$ in a smooth way to span the tangent space to each 2-surface $S_{uv} = \mathcal{N}_u \cap \mathcal{N}_v'$, to be null and to be normalized via $m.\overline{m} = -1$. Note that there is **spin freedom**, $m \mapsto e^{i\theta} m$ with θ real, available at each point. This is best used to simplify particular problems. All the remaining conditions for $(l, n, m, \overline{m})$ to form a NP tetrad are automatically satisfied.

On the generators of each $\mathcal{N}_u$ only v is varying. This means that the vector field l is given by

$$l = Q \partial / \partial v, \qquad (4.4.25)$$

where Q is a function of position. Since the vector field n is tangent to the generators of each $\mathcal{N}_v'$ and $n^\mu \nabla_\mu u = 1$ we have

$$n = \partial / \partial u + C^A \partial / \partial x^A, \qquad (4.4.26)$$

where the real C^A are functions of position. Note that by construction the x^A do not vary on the generators of $\mathcal{N}_0'$. Thus $C^A = 0$

on $\mathcal{N}_0'$. Finally since m, $\bar{m}$ span the tangent space of each $\mathcal{S}_{uv}$ we have

$$m = P^A \partial / \partial x^A. \qquad (4.4.27)$$

Note that the condition $m.\bar{m} = -1$ implies that there are only 3 real functions of position involved in the P^A. These plus the real Q, C^A give 6 real scalar fields describing the metric. With this choice of tetrad we automatically have

$$\kappa = \nu = \epsilon = \rho - \bar{\rho} = \mu - \bar{\mu} = \tau - \bar{\alpha} - \beta = 0 \qquad \text{in } \mathcal{M}. \quad (4.4.28)$$

In addition we have

$$\gamma - \bar{\gamma} = 0 \qquad \text{on } \mathcal{N}_0'. \qquad (4.4.29)$$

Apart from the initial choice of Q and the spin freedom, the coordinate and tetrad gauge freedom has been used up.

When we come to examine the characteristic initial value problem in detail, a remarkable simplification occurs. This was first noted by Sachs, 1962. Let us first consider the unconstrained Maxwell system given by equations (4.1.4) or (4.1.7–8) on a given background. As we saw at the end of section 4.3, ϕ_0, ϕ_1 form the u-variables on $\mathcal{N}_0$ and can be specified freely there. In the third equation (4.1.8), $\bar{\delta}\phi_1$ is already known and so this is an **ordinary differential equation** (along the generators of $\mathcal{N}_0$) for ϕ_2. Similarly the fourth one is an ordinary differential equation for ϕ_3. In both cases the initial data is given on $\mathcal{S} = \mathcal{N}_0 \cap \mathcal{N}_0'$. These equations will be coupled in general. In this case they are actually linear, and so can be solved analytically. There is no essential change for the constrained system where $\phi_3 = \phi_1$ and the u-variable is ϕ_0. Now the fourth equation (4.1.8) can be regarded as an ordinary differential equation for ϕ_1. Since $\kappa = 0$ by construction, this equation does not involve ϕ_2, and so can be solved analytically. Then the third equation (4.1.8) can be solved for ϕ_2 on $\mathcal{N}_0$ with ϕ_0, ϕ_1 known there. Of course identical considerations apply on $\mathcal{N}_0'$.

When we move to the gravitational case very little changes in principle, but the number of variables is considerably enlarged, and some organization is needed to keep track of what is happening. First we deal with the freedom of choice in the hypersurface labelling. This is equivalent to giving Q freely on both $\mathcal{N}_0$ and $\mathcal{N}_0'$. In addition we have $C^A = 0$ on $\mathcal{N}_0'$.

On $\mathcal{N}_0$ the physically important data are the matter and radiation fluxes crossing the hypersurface transversely. On physical grounds one would expect to be able to choose them freely, and the NP scalars describing them are Φ_{00} and Ψ_0 respectively. These constitute the u-variables on $\mathcal{N}_0$. Analogous considerations apply on $\mathcal{N}_0'$, where the corresponding quantities are Φ_{22} and Ψ_4. All four of these quantities are specified on S the intersection of the two initial hypersurfaces.

For the standard Cauchy problem one has to specify both the intrinsic and extrinsic curvatures (which however are subject to constraints). The intrinsic curvature is the internal metric, and indeed we have to give this freely on S by choosing P^A. The extrinsic curvature describes how the initial surface is embedded in the spacetime, i.e., the rates of change of the divergence and shear of the normal covector field. Here we need to know how S is embedded in each of $\mathcal{N}_0, \mathcal{N}_0'$, and this can be specified by giving ρ, σ, μ, λ freely on S. The final data required on S concerns the conformal factor. We may specify Ω and both s_0 and s_2 on S. All of the remaining dependent variables may be determined using only differentiation and algebraic processes. This data can usually be determined analytically.

On $\mathcal{N}_0$ we deal with the conformal factor first. Note that on $\mathcal{N}_0$, $D = Q\partial/\partial v$ is known. The equations governing Ω and its v-derivative are

$$D\Omega = s_0, \quad Ds_0 = -\Omega\Phi_{00}. \tag{4.4.30}$$

This is a first order linear system with data given on S. There is no difficulty in solving this for Ω and s_0 on $\mathcal{N}_0$, and indeed this can usually be done analytically. Next we consider ρ and σ on $\mathcal{N}_0$. Here the governing equations are

$$D\rho = \rho^2 + \sigma\bar{\sigma} + \Phi_{00}, \quad D\sigma = 2\rho\sigma + \Psi_0, \tag{4.4.31}$$

with initial data given on S. These are of course nonlinear but the nonlinearity is of a well-defined kind. Adopting the definitions

$$Y = \begin{pmatrix} \rho & \sigma \\ \bar{\sigma} & \rho \end{pmatrix}, F = \begin{pmatrix} \Phi_{00} \\ \Psi_0 \end{pmatrix},$$

we have

$$DY = Y^2 + F, \tag{4.4.32}$$

a **Riccati** equation for the Hermitian matrix Y. (See also section 2.7.) Assuming this system has been solved on $\mathcal{N}_0$ all the remaining equations there are linear, and when solved in the right order (see Stewart and Friedrich, 1982, for details) are ordinary. Similar considerations apply on $\mathcal{N}_0'$.

The simplicity of the "initial value problem" has been exploited by Isaacson et al., 1983, to explore the singular case where S and $\mathcal{N}_0'$ degenerate to a point, and $\mathcal{N}_0$ is a null cone. Evolution off the initial surfaces is however non-trivial.

4.5 Lagrangian and Legendrian maps

In section 4.3 it was implicitly assumed that the characteristic hypersurfaces were smooth. However these hypersurfaces were determined as solutions of the system (4.3.5) with initial conditions (4.3.6) in a "phase space" T^*M. While the solution is certainly smooth in phase space we are interested in its projection into spacetime M. Unfortunately the projection map can become singular, so that the (projection of the) characteristic hypersurfaces can develop singularities or **caustics**. In the remainder of this chapter we develop the general theory of caustics in general relativity, based on Friedrich and Stewart, 1983. This work relies heavily on the results developed by Arnol'd and his pupils. Specialized references are given in the work cited, and a good introductory reference is Arnol'd, 1978. An exhaustive detailed account can be found in Arnol'd et al., 1985. However the following account is designed to be self-contained. In this section we set up some basic concepts.

Let M be a n-manifold with local coordinates (q^i). Let T^*M be its cotangent bundle with local coordinates (q^i, p_i), projection map $\pi : T^*M \to M$, $\pi(q,p) = q$. The point (q,p) corresponds to the covector $\alpha = p_i dq^i$ at q. Finally let $PT^*M = (T^*M \setminus \{0\})/R^*$ be the projectified cotangent bundle, obtained from the set of nonvanishing covectors by factoring out the nonvanishing real numbers, i.e., if $\alpha \neq 0$ we identify α and $x\alpha$ for real $x \neq 0$. Each fibre of PT^*M is a $(n-1)$-dimensional real projective space. We may define two projections $\pi' : PT^*M \to M$, $\pi'' : T^*M \setminus \{0\} \to PT^*M$, by requiring $\pi'(\pi''(a)) = \pi(a)$ for all $a \in T^*M$, $a \neq 0$. In a neighbourhood of some nonzero $a \in T^*M$, at least one coordinate $p_i \neq 0$ and without loss of generality we assume $p_n(a) \neq 0$. We can now define an *affine coordinate system* (q^k, z, p_k), $k = 1, \ldots, n-1$, for PT^*M in a neighbourhood of $\pi''(a)$ by setting $p_n = 1$, $q^n = z$.

A standard 1-form $\tilde{\kappa}$ on T^*M, i.e., an element of T^*T^*M, can be defined in a natural way as follows. Let $X \in TT^*M$ be an arbitrary vector field on T^*M so that $\pi_* X$ is a vector field on M. Let $a \in T^*M$ and let $p(a)$ be the 1-form in M at $\pi(a)$. Define $\tilde{\kappa}$ at a by

$$\tilde{\kappa}(X) = p(a)(\pi_* X), \qquad (4.5.1)$$

for all X. It is a simple exercise to verify the coordinate representation

$$\tilde{\kappa} = p_i \, dq^i + 0 \, dp_i. \qquad (4.5.2)$$

The standard 2-form $\tilde{\omega}$ on T^*M is

$$\tilde{\omega} = -d\tilde{\kappa}. \qquad (4.5.3)$$

In coordinates we have

$$\tilde{\omega} = dq^i \wedge dp_i. \qquad (4.5.4)$$

It is obvious from (4.5.3) that $\tilde{\omega}$ is *closed*, i.e., $d\tilde{\omega} = 0$, and from (4.5.4) that $\tilde{\omega}$ is *non-degenerate*, i.e., for any $X \in TT^*M$, $X \neq 0$, $\exists Y \in TT^*M$ such that $\tilde{\omega}(X, Y) \neq 0$. A closed non-degenerate differential 2-form on an (even-dimensional) manifold is usually called a *symplectic structure*. (See also section 2.2.) There is a converse result: if a manifold M admits a symplectic structure $\tilde{\omega}$ then a local coordinate chart (q^i, p_i) can be found such that (4.5.4) holds, i.e., every symplectic structure is diffeomorphic to (4.5.4). This is *Darboux's theorem*, see e.g., Arnol'd, 1978, section 43B. Coordinate transformations $(q, p) \rightarrow (Q, P)$ which preserve the form (4.5.4), i.e., $\tilde{\omega} = dQ^i \wedge dP_i$ are called *canonical transformations* in classical mechanics. (In general however they do not preserve the property of being adapted to the bundle structure.)

The *null spaces* of $\tilde{\kappa}$ are the subspace of vectors $X \in TT^*M$ such that $\tilde{\kappa}(X) = 0$. $\tilde{\kappa}$ is not invariant under multiplication in the fibres of $T * M$, for it transforms homogeneously. However the null spaces of $\tilde{\kappa}$ are invariant, and are projected by π''_* onto hypersurfaces in $TPT * M$. This field of 'contact hypersurfaces' defines a *contact field* κ which annihilates vectors in the contact hypersurfaces. Evidently in affine coordinates we may take

$$\kappa = p_k \, dq^k + dz. \qquad (4.5.5)$$

Now $\omega = d\kappa$ must be degenerate since PT^*M has odd dimension. However the restriction to the contact hypersurfaces is non-degenerate and κ defines a natural *contact structure* on PT^*M. (For an elegant introduction to contact structures see Arnol'd, 1978, appendix 4.) There is a Darboux theorem which states that for every odd-dimensional manifold carrying contact structure, the contact field can be expressed in the form (4.5.5), i.e., all such manifolds are locally diffeomorphic. Note that although we have obtained these structures by projection from T^*M one can start from the odd-dimensional manifold PT^*M, erect a bundle structure on it with fibres diffeomorphic to R, and construct the symplectic structure from the contact one. This process is called *symplectification*.

Let $\tilde{N}$ be a smooth submanifold of T^*M with embedding $\tilde{\iota} :$ $\tilde{N} \rightarrow T^*M$. The pull-back of $\tilde{\omega}$ defines a 2-form on $\tilde{N}$. If this vanishes

$$\tilde{\iota}^* \tilde{\omega} = 0, \qquad (4.5.6)$$

then $\tilde{N}$ is said to be *isotropic*. The maximum dimension of an isotropic submanifold is n, the dimension of M, and isotropic manifolds of maximal dimension are **Lagrangian manifolds**. The projection $\bar{\pi} = \pi \circ \bar{\imath} : \tilde{N} \to M$ is a **Lagrangian map**. Trivial examples of Lagrangian manifolds in T^*M are the base manifold and the fibres. (For a non-trivial example let $M = R^n$, let V be a smooth submanifold of M and let $\tilde{N}$ be the space of vectors normal to V. $\tilde{N}$ has dimension n. Using the usual Euclidean conventions we identify $\tilde{N}$ with a subspace $N \subset T^*M$. Clearly $\tilde{\kappa} = 0$ on N and so N is a Lagrangian manifold.) More generally if $\tilde{N}$ is a Lagrangian manifold (4.5.3), (4.5.6) imply that

$$\bar{\imath}^* \tilde{\kappa} = d\tilde{u}, \qquad (4.5.7)$$

for some scalar field $\tilde{u}$ on $\tilde{N}$.

Now, provisionally, assume that $\bar{\pi}$ is a diffeomorphism. Then there exists a function $u(q)$ on M such that $\tilde{u} = \bar{\pi}^*u$. Equations (4.5.7), (4.5.2) then imply that

$$\tilde{N} = \left\{ (q^i, p_i) : \quad p_i = \partial u / \partial q^i \right\}, \qquad (4.5.8)$$

for some $u : M \to R$. Conversely if $\tilde{N}$ is given by (4.5.8) it has dimension n, $\bar{\pi}^*\tilde{\kappa} = d\tilde{u}$ and $\tilde{\omega} = 0$ so that $\tilde{N}$ is a Lagrangian manifold. In general $\bar{\pi}$ will not be a diffeomorphism; suppose it has rank $k < n$. Then there exists a subset J of $\{1, \ldots, n\}$ with k elements and its complement I, such that the image of $\bar{\pi}$ is the submanifold of M spanned by the (q^j) with $j \in J$ (which we henceforth abbreviate to q^J).

Next consider a canonical transformation

$$Q^J = q^J, \quad P_J = p_J, \qquad Q^I = p_I, \quad P_I = -q^I,$$

which obviously leaves the symplectic structure and the concept of a Lagrangian manifold unchanged. Rewrite $u(Q)$ as $u(Q) = -S(q^J, P_I)$. Then (4.5.8) becomes

$$\tilde{N} = \left\{ (q, p) : \quad q^I = \partial S / \partial p_I, \ p_J = -\partial S / \partial q^J \right\}. \qquad (4.5.9)$$

This is the coordinate representation of a general Lagrangian manifold with **generating function** $S(q^J, p_I)$. The apparent asymmetry between (4.5.8) and (4.5.9) is removed by defining a *generating family of functions*

$$F(q,p_I) = -S(q^J,p_I) + q^I p_I, \qquad (4.5.10)$$

in a $(2n - k)$-dimensional space with coordinates (q,p_I). An n-dimensional submanifold is given by imposing the constraints

$$\partial F/\partial p_I = 0, \qquad (4.5.11)$$

and this can be embedded into an open set of the Lagrangian manifold by setting

$$p_i = \partial F/\partial q^i, \qquad 1 \le i \le n. \qquad (4.5.12)$$

Conversely any smooth function $F(q,p_I)$ such that the rank of (4.5.11) is $n - k$ defines a Lagrangian manifold via (4.5.11), (4.5.12).

There is an analogous theory in PT^*M. A submanifold N of PT^*M is an *integral manifold* of the field of contact hypersurfaces if

$$\iota^*\kappa = 0, \qquad (4.5.13)$$

where $\iota : N \to PT^*M$ is the embedding. If N has maximal dimension for this property, i.e., $n - 1$ then N is a **Legendrian manifold**. The projection $\hat{\pi} = \pi' \circ \iota : N \to M$ is a **Legendrian map**. An important example will appear in the next section. One can guess the general coordinate representation corresponding to (4.5.9) for Lagrangian manifolds. Suppose I and J are complementary subsets of $\{1,2,\ldots,n\}$ and let (q^J,p_I) be a coordinate chart for N. Let $S(q^J,p_I)$ be an arbitrary smooth function of its arguments. Define

$$q^I = \partial S/\partial p_I, \qquad p_J = -\partial S/\partial q^J,$$

in analogy with (4.5.9). In order for (4.5.13) to hold $p_k dq^k$ must be exact. Now

$$
\begin{aligned}
p_k dq^k &= p_I dq^I + p_J dq^J \\
&= p_I d\left(\frac{\partial S}{\partial p_I}\right) - \frac{\partial S}{\partial q^J} dq^J \\
&= d\left(p_I \frac{\partial S}{\partial p_I}\right) - \frac{\partial S}{\partial p_I} dp^I - \frac{\partial S}{\partial q^J} dq^J \\
&= d\left(p_I \frac{\partial S}{\partial p_I} - S\right).
\end{aligned}
$$

Thus a Legendrian manifold is given in affine coordinates by

$$
N = \{(q, z, p) : q^I = \partial S/\partial p_I, z = S - \partial S/\partial p_I, p_J = -\partial S/\partial q^J\}.
\tag{4.5.14}
$$

To construct a generating family consider

$$
F(q, z, p_I) = z + q^I p_I - S(q^J, p_I),
\tag{4.5.15}
$$

defined on a $(2n - k)$-dimensional space with coordinates q, z, p_I. An $(n - 1)$-dimensional submanifold can be obtained by imposing the $(n - k) + 1$ constraints

$$
F = 0, \qquad \partial F/\partial p_I = 0.
\tag{4.5.16}
$$

This can be embedded into an open set of the Legendrian manifold via

$$
p_k = \partial F/\partial q^k, \qquad 1 \le k \le n - 1.
\tag{4.5.17}
$$

Conversely any smooth function $F(q, z, p_I)$ such that the rank of (4.5.16) is $n - k + 1$ defines a Legendrian manifold via (4.5.16), (4.5.17).

4.6 Wavefronts and wavefront families

We now impose a metric g on M with signature $(+ - - - \cdots)$. In a relativistic theory we expect the characteristic surfaces of section 4.3 to be null hypersurfaces with respect to g, and this is borne out by the discussion in section 4.4. We therefore review the construction given in section 4.3 in terms of Lagrangian and Legendrian maps. For convenience the traditional coordinate chart (x^i) will be denoted (q^i), maintaining consistency with section 4.5.

g defines a Hamiltonian function on $T * M$ via

$$H(q,p) = \tfrac{1}{2}g^{ik}(q)p_i p_k.$$

Let $\tilde{H}$ be the hypersurface of non-vanishing null covectors in T^*M,

$$\tilde{H} = \{(q,p) \in T^*M : \quad p \neq 0, \ H(q,p) = 0\}. \qquad (4.6.1)$$

Let ω be the symplectic form on T^*M. The Hamiltonian vector field $X_H \in TT^*M$ is defined as follows. Let $Y \in TT^*M$ be an arbitrary vector field and require

$$\omega(X_H, Y) = dH(Y), \qquad \forall Y \in TT^*M. \qquad (4.6.2)$$

In a coordinate-induced basis

$$X_H = \left(\frac{\partial H}{\partial p_i}, -\frac{\partial H}{\partial q^i}\right). \qquad (4.6.3)$$

The integral curves of X_H define the Hamiltonian flow lines in T^*M. Their projection into M produces the null geodesics in M.

Now suppose $\tilde{M} \subset \tilde{H}$ is a Lagrangian submanifold, and let Y be a vector field in $T\tilde{M}$. Clearly $Y \in T\tilde{H}$ and so (4.6.1) implies $dH(Y) = Y(H) = 0$. Now (4.6.2) implies that ω is annihilated by the vector space spanned by all such Y and X_H. Since by definition $T\tilde{M}$ is the space of maximal dimension with this property we deduce that $X_H \in T\tilde{M}$. Thus if $\tilde{Q}$ is a hypersurface of $\tilde{M}$ transverse to X_H we may regard $\tilde{M}$ as being generated from $\tilde{Q}$ by dragging $\tilde{Q}$ along the Hamiltonian flow lines. If as in the previous section $\tilde{\kappa}$ denotes the canonical 1-form then using (4.6.3), $\tilde{\kappa}(X_H) = p_i \partial H/\partial p_i = 2H = 0$ on $\tilde{H}$. Now within $\tilde{M}$ we may use (4.5.7) to set $\tilde{\kappa} = d\tilde{u}$, for some function $\tilde{u}$. Thus $\tilde{u}$ is constant along the Hamiltonian flow lines and so is determined by its values on $\tilde{Q}$.

We shall assume that the Lagrangian map $\tilde{\pi}$ is regular at $\tilde{Q}$, i.e., $Q = \tilde{\pi}\tilde{Q}$ is a smooth hypersurface in M. In particular this means that $\tilde{\kappa} = d\tilde{u} \neq 0$ on $\tilde{Q}$. It is straightforward to verify that $\mathcal{L}_{X_H}\tilde{\kappa} = 0$ in $\tilde{M}$, from which we may deduce that $\tilde{\kappa} = d\tilde{u}$ vanishes nowhere in $\tilde{M}$. Now X_H is transverse to $\tilde{Q}$ and $\tilde{\pi}_* X_H{}^i = \partial H/\partial p_i \neq 0$ from the definition of $\tilde{H}$. Thus $\tilde{\pi}$ defines a diffeomorphism of a neighbourhood $\tilde{U}$ of $\tilde{Q} \subset \tilde{M}$ to a neighbourhood U of $Q \subset M$. We define a scalar field u on U via

$$u = \tilde{u} \circ \tilde{\pi}^{-1}. \tag{4.6.4}$$

It now follows that in U,

$$g^{ik}u_{,i}u_{,k} = 0, \qquad u_{,i} \neq 0, \tag{4.6.5}$$

i.e., u is a *null coordinate* on U. There is a converse construction. Suppose we choose a smooth hypersurface Q in M, and a non-vanishing null covector field on Q which is the pull-back of the differential of a function defined in a neighbourhood of Q. This null form maps Q onto a smooth $(n-1)$-dimensional surface $\tilde{Q}$ in $T^*M \cap \tilde{H}$. By dragging $\tilde{Q}$ along the Hamiltonian flow we may generate a Lagrangian manifold.

For each real number v we define

$$\tilde{N}_v = \left\{ (q,p) \in \tilde{M} : \quad \tilde{u}(q,p) = v \right\}. \tag{4.6.6}$$

In the previous paragraph we saw that $\tilde{\kappa} = d\tilde{u}$ was non-vanishing in $\tilde{M}$ and so (4.6.6) defines smooth $(n-1)$-dimensional submanifolds of $\tilde{M}$ with smooth inclusion map $\iota_v : \tilde{N} \to \tilde{M}$. Clearly on $\tilde{N}_v$

$$\iota_v{}^* \tilde{\kappa} = 0. \tag{4.6.7}$$

Using the projection π'' of section 5 we may map $\tilde{N}_v$ to a Legendrian manifold $N_v \subset PT^*M$, and note also that it is contained in the hypersurface of PT^*M defined by $H = 0$. Using the Lagrangian map $\tilde{\pi}$ we may also map $\tilde{N}_v$ into the subset of $U \subset M$ defined by $u = v$. Close to Q this will be a smooth null hypersurface, but further away there is no reason why $\tilde{\pi}$ should be a diffeomorphism. (Equivalent statements hold for the Legendrian maps $\hat{\pi}_v = \pi \circ \iota_v$.)

Let Y be a nowhere vanishing vector field on $\tilde{M}$ which is Lie-propagated by X_H, and is tangent to $\tilde{Q}$. Under projection by the Lagrangian map $\bar{\pi}$, Y becomes a Jacobi field along the null geodesics in M obtained by projecting the Hamiltonian flow lines. At a typical point $a \in \tilde{Q}$ one will have $\tilde{\kappa}(Y) \neq 0$. Since both $\tilde{\kappa}$ and Y are Lie-propagated along the flow lines, this relation is maintained along the flow lines. Thus Y is transverse to the hypersurface $\tilde{N}_v$ containing a, and the projection of Y to TM vanishes nowhere along the corresponding null geodesics. This is the regular case.

At some points $a \in \tilde{Q}$ one might expect $\tilde{\kappa}(Y) = 0$. This will then hold at all points of the flow line through a. Thus Y is tangent to the $\tilde{N}_v$ containing a. Along the flow line there may be points at which Y becomes tangent to the fibres of T^*M, where the projection of Y vanishes. These points generate the **caustic set** of $\bar{\pi}$ and $\hat{\pi}_v$ and they are mapped onto points of the null geodesics through $\hat{\pi}_v(a)$, projections of the flow line through a, which are conjugate with respect to the "initial surface" $Q \cap \bar{\pi}(\tilde{N}_v)$. Since the vector fields Y together with X_H span $T\tilde{M}$ it follows that the Lagrangian map $\bar{\pi}$ (the Legendrian map $\hat{\pi}_v$) drops in rank where $\tilde{N}_v$ becomes tangent to the fibres of T^*M (where N_v becomes tangent to the fibres of PT^*M), and that it cannot drop in rank by more than $n-2$. Near the caustic the projections of the flow lines may or may not intersect each other. However the null coordinate u cannot be extended to the caustic and the image of the Legendrian map $\hat{\pi}_v$ no longer provides a smooth hypersurface.

4.7 Caustics in Minkowski spacetime

It is of some interest to identify the different stable caustics possible in spacetime. Since this is essentially a local problem we restrict attention to Minkowski spacetime. We must now make precise what we mean by 'different'. A transformation of PT^*M which preserves the fibres and contact structure is a **Legendrian equivalence**. If it maps one Legendrian submanifold N_1 to another N_2, then N_1, N_2 are **Legendre equivalent**. We are looking only for the different equivalence classes. The definition of 'stability' is somewhat technical. The germ of a Legendrian manifold N at the origin o is *stable* if for any Legendrian manifold N' close enough to N (in the Whitney C^∞ topology) there exists a point o' close to the origin such that the germ of N at o is Legendre equivalent to the germ of N' at o', Arnol'd, 1978, Arnol'd et al., 1985.

The classification of the stable caustics in Minkowski spacetime, up to Legendre equivalence, was given by Friedrich and Stewart, 1983. (See also Arnol'd et al., 1985.) However most of the results of their somewhat technical discussion can be reviewed more simply using NP techniques.

Let $\mathcal{N}$ be a null hypersurface emanating from a spacelike 2-surface S. Let $(z, \bar{z})$ be (complex) coordinates on S and let v be an affine parameter along the geodesic generators of $\mathcal{N}$. We extend $(z, \bar{z})$ onto $\mathcal{N}$ by requiring them to be constant along the generators of $\mathcal{N}$. This construction of intrinsic coordinates $(v, z, \bar{z})$ on $\mathcal{N}$ has been used earlier in sections 4.4 and 4.6. Suppose that we could extend this to a coordinate system $(u, v, z, \bar{z})$ in M where $\mathcal{N}$ is $u = 0$. From the previous section we deduce that a caustic occurs when the Jacobian matrix of $(u, v, z, \bar{z})$ with respect to Cartesian coordinates x^a loses rank. (It is shown in Friedrich and Stewart, 1983, that the maximum corank is two.) This construction is both difficult to carry out and requires the extraneous coordinate u, so that we look instead for a more intrinsic characterisation of caustics.

Suppose we attempt to construct a NP null tetrad with l tangent to the generators and $(l, m, \overline{m})$ spanning $T\mathcal{N}$. This tetrad has considerable freedom, for besides the boosts and spins we may make a null rotation about l, see appendix B. Indeed a choice of the final vector n is the choice of a particular null rotation. We therefore ask what properties are independent of such null rotations. The only connection coefficient which is manifestly invariant under null rotations about l is κ. Here $\kappa = 0$ since l is geodesic. This implies, see appendix B, that ρ, σ and ϵ are invariant. As in section 4.4 affine parametrization of l implies that we may set $\epsilon - \bar{\epsilon} = 0$, $\rho - \bar{\rho} = 0$ and by means of a boost we may set $\epsilon = 0$.

Thus the only invariant intrinsic information is contained in ρ and σ. The "field" equations (a) and (b) of appendix B imply

$$D\rho = \rho^2 + \sigma\bar{\sigma}, \qquad D\sigma = 2\rho\sigma. \qquad (4.7.1)$$

Using a spin we may ensure that σ is real. We expect a caustic to occur at points where ρ and σ become singular. In fact this is closely related to the breakdown of coordinate systems described in the previous paragraph. For in section 2.7 it was shown that the complex coordinate z can be propagated by

$$Dz = -\rho z - \sigma\bar{z}. \qquad (4.7.2)$$

Since l remains non-singular a caustic occurs precisely when the $(z,\bar{z})$ coordinates become singular, i.e., when ρ and σ develop singularities.

Next let $w_{\pm} = \rho \pm \sigma$, so that (4.7.1) implies

$$Dw_{\pm} = w_{\pm}^2. \qquad (4.7.3)$$

Setting $D = \partial/\partial v$, with $v = 0$, $w_{\pm} = w_{\pm o}$ on S we have

$$w_{\pm} = \frac{1}{w_{\pm o}^{-1} - v}. \qquad (4.7.4)$$

Thus a knowledge of $w_{\pm o}(z,\bar{z})$ is sufficient to locate and analyse the caustics. Note that

$$\rho \sim \sigma \sim \frac{\frac{1}{2}}{w_{+o}^{-1}(z,\bar{z}) - v} \qquad \text{as } v \to w_{+o}^{-1}, \qquad (4.7.5)$$

with a similar representation for the other case. Let us now readjust the origin of v placing it in a neighbourhood of the caustic.

Consider first the case where there is a symmetry, so that w_{+o} is a function of a single variable, say $w_{+o}^{-1} = f(x)$. Thus a caustic corresponds to a zero of $f(x) - v$. A generic zero of $f(x)$ at $x = 0$ can occur in one of two ways. The simplest occurs if f is locally a linear function

$$f(x) = \lambda x + O(x^2), \qquad \lambda \neq 0. \qquad (4.7.6)$$

Thus $\rho \sim \sigma \sim \frac{1}{2}/(\lambda x - v)$ near $x = 0$, $v = 0$. Arnol'd calls this an A_2-*caustic*. Using (4.7.2), (4.7.4) it is easy to see that the projection of $\mathcal{N}$ onto the $z, \bar{z}$-plane develops a cusp. An explicit example has been given by Friedrich and Stewart, 1983. It has a generating function

$$S(q^2, q^3, p_1) = \epsilon p_1{}^3 + \sqrt{1 - p_1{}^2}\, q^3, \qquad (4.7.7)$$

where $\epsilon = \pm 1$. Using Cartesian coordinates x^a, setting $l^0 = 1$, $s = \sin\theta = -p_1$, and $c = \sqrt{1 - s^2}$, (4.5.14) implies that $\mathcal{N}$ is given by

$$x^a = (2\epsilon s^3 + v, 3\epsilon s^2 + sv, cv, q^3). \qquad (4.7.8)$$

$T\mathcal{N}$ is spanned by

$$l^a = \frac{\partial x^a}{\partial v}, \quad e^a = \frac{\partial x^a}{\partial \theta}, \quad f^a = \frac{\partial x^a}{\partial q^3}.$$

f^a is a spacelike unit vector. However $e^a e_a = -(v + 6\epsilon sc^2)^2$, and so the caustic occurs when $v = -6\epsilon s(1 + O(s^3))$. A spacetime picture of the A_2-caustic is shown in fig. 4.7.1. (This and the other figures in this section are adapted from Friedrich and Stewart, 1983, which also contains a more detailed discussion.) Two points should be emphasised. Firstly this is a representation of the A_2-caustic only in a neighbourhood of $s = 0$, $v = 0$. Secondly every other A_2-caustic in Minkowski spacetime is locally diffeomorphic to this one (up to Legendre equivalence).

The only other generic zero of $f(x)$ is a quadratic one

$$f(x) = \lambda x^2 + O(x^3), \qquad (4.7.9)$$

and without loss of generality we may set $\lambda = 1$. For $v > 0$ there are two A_2-lines at $x = \pm\sqrt{v}$ and these merge as $v \to 0$. For $v < 0$ there is no caustic. Continuity implies that besides the two cusp lines there must be a self-intersection line in $\mathcal{N}$ and it is easy to see that all three are tangent at the A_3-*caustic* $x = v = 0$. An explicit example is given by the generating function

$$S(q^2, q^3, p_1) = \epsilon p_1{}^4 + \sqrt{1 - p_1{}^2}\, q^3. \qquad (4.7.10)$$

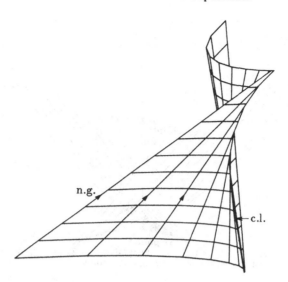

Fig. 4.7.1   The intersection of the wavefront $\mathcal{N}$ with $q^3 = constant$ for an A_2-caustic. n.g. denotes a null generator and c.l. the caustic line. In this and subsequent pictures the computer programme has represented the curved surfaces by a set of plane quadrilaterals. In some cases the intersection lines, correct for the quadrilaterals, appear a little uneven. The reader should interpolate mentally a smooth curve.

Here $\mathcal{N}$ is given in Cartesian coordinates by

$$x^a = (3\epsilon s^4 + v, 4\epsilon s^3 + sv, cv, q^3), \qquad (4.7.11)$$

and $e^a e_a = -(v + 12\epsilon s^2 c^2)^2$ so that the caustic is given locally by $v = -12\epsilon s^2 + O(s^4)$. A spacetime picture is depicted in fig. 4.7.2. Again this is only a valid representation of the caustic in the neighbourhood of $v = s = 0$, and every A_3-caustic in Minkowski spacetime is diffeomorphic to this one (up to Legendrian equivalence).

Once the spacetime symmetry is removed two additional caustics can occur. Consider first a caustic at

$$v = f(x, y) = \lambda x(\mu x^2 - y) + O(x^4, x^2 y, y^2), \qquad (4.7.12)$$

where without loss of generality $\lambda = \mu = 1$. For $y < 0$ there is a single A_2-caustic line, but for $y > 0$ there are two additional

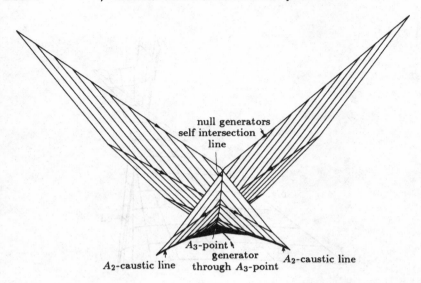

Fig. 4.7.2   The intersection of wavefront $\mathcal{N}$ with $q^3 = constant$ for an A_3-caustic. There is precisely one generator through the A_3-point. All other generators pass through the self-intersection line before hitting one of the A_2-caustic lines. These two lines and the self-intersection line touch at the A_3-point.

caustic lines at $x = \pm\sqrt{y}$, which emerge from the A_4-**caustic** at $v = x = y = 0$. (4.7.12) is of course the equation of the "Whitney pleat surface". Thus we have an A_3-caustic emerging from an A_2-caustic. There are 3 A_2-lines and 3 self-intersection lines all tangent at the A_4-point. A specific example is given by

$$S(q^2, q^3, p_1) = \epsilon p_1{}^5 + (1 - p_1{}^2)q^2 + p_1\sqrt{1 - p_1{}^2}\, q^3. \qquad (4.7.13)$$

Setting $q^2 = 0$, $s = \sin\theta = -p_1$, $c = \cos\theta$ and $q^3 = cy$ we may represent $\mathcal{N}$ with respect to a Cartesian coordinate system by

$$x^a = (4\epsilon s^5 + s^3 y + v,\ 5\epsilon s^4 - (1 - 2s^2)y + sv,\ -c^2 v,\ c(y + sv)). \quad (4.7.14)$$

$\mathcal{N}$ is spanned by

$$l^a = \frac{\partial x^a}{\partial v}, \quad e^a = \frac{\partial x^a}{\partial \theta}, \quad f^a = \frac{\partial x^a}{\partial f}.$$

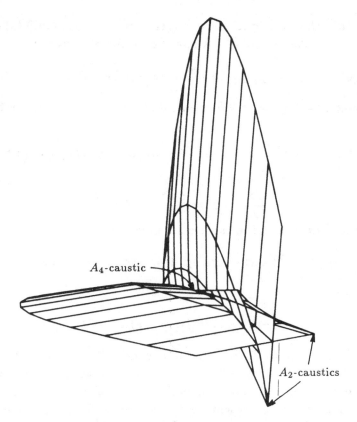

Fig. 4.7.3   The A_4 caustic at $t = x^0 = 0$. There are three A_2-caustic lines and three self-intersection lines all tangent at the A_4-point. At earlier and later times these split into a single A_2-line and and a pair of A_2-lines ending tangent at an A_3-point.

$f^a f_a = -c^4(1 + c^2)$, and so f^a remains regular. However e^a becomes null when

$$v = -(20\epsilon s^3 c^2 + (3 - 2s^2)ys)/(1 + c^2)$$
$$\approx -\tfrac{3}{2}\left(y + \tfrac{20}{3}s^2\right),$$

which is diffeomorphic to (4.7.12). The A_4-caustic is illustrated in fig. 4.7.3.

The final case to be considered is the D_4-***caustic***, where the rank of the projection map drops by two rather than one. This means that the caustic set is no longer a smooth submanifold, but bifurcates at the D_4-point. A specific example is given by

$$S(q^3, p_1, p_2) = \epsilon' p_1{}^2 p_2 + \epsilon p_2{}^3 + \sqrt{1 - p_1{}^2 - p_2{}^2}\, q^3. \qquad (4.7.15)$$

Choosing $\epsilon' = 1$, $a = -p_1$, $b = -p_2$ and $c = \sqrt{1 - a^2 - b^2}$ we find $\mathcal{N}$ given by

$$x^a = (2a^3 + 2\epsilon ab^2 + v, -3a^2 - \epsilon b^2 + av, -2\epsilon ab + bv, cv). \quad (4.7.16)$$

Now

$$l^a = \frac{\partial x^a}{\partial v}, \quad e^a = \frac{\partial x^a}{\partial a}, \quad f^a = \frac{\partial x^a}{\partial b}$$

span $T \times \mathcal{N}$ except at the caustic set given by

$$v^2 + 2a[3(1 - a^2) + \epsilon(1 - 3b^2)]v + 4c^2(3\epsilon a^2 - b^2) = 0. \quad (4.7.17)$$

For small a, b,

$$v \approx -(3 + \epsilon)a \pm 2\sqrt{(3 - \epsilon)^2 a^2 + b^2},$$

which is diffeomorphic to a double cone. The two cases are sketched in figs. (4.7.4), (4.7.5). It is both fortunate and surprising that more complicated stable caustics do not occur.

Caustics obviously present computational difficulties when integrating the characteristic initial value problem although it is obvious, in principle, how to integrate through the caustic. They do however present fundamental issues. Consider first the A_2-caustic given by (4.7.8), and consider the separation between neighbouring points $v = 0$, $s = s_o(1 + \alpha)$ and $v = v_o$, $s = s_o$ where α and v_o are small. It is easy to show that

$$ds^2 = -6\alpha^2 s_o{}^3(v_o + 6\epsilon s_o c_o^2) + O(\alpha^3). \qquad (4.7.18)$$

Now if both points are on the same side (with respect to v) of the caustic the separation is either spacelike ($\alpha \neq 0$) or null ($\alpha = 0$). However if the caustic passes between them, then the separation becomes timelike ($\alpha \neq 0$). Therefore initial data for the characteristic initial value problem can only be given on S and along the geodesics until the A_2-caustic is reached. Subsequent points

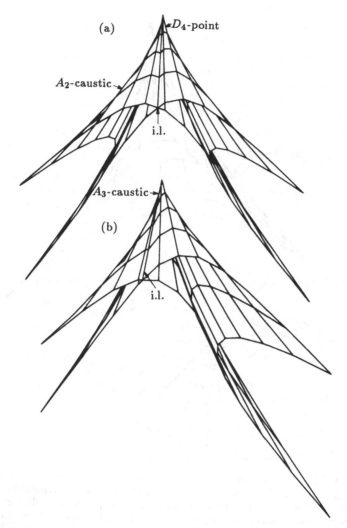

Fig. 4.7.4 The D_4 wavefront when $\epsilon = 1$. (a) The intersection of the wavefront with $t = x^0 = 0$. There are four A_2-caustic lines, which touch the one self- intersection line, i.l. at a D_4 caustic point. (b) The intersection of the wavefront with $t = constant > 0$. The upper two A_2-lines touch the self-intersection line at an A_3- point. The lower two A_2-lines touch, but do not meet the other A_2-lines or the self-intersection line. For $t = constant < 0$ the rôle of the two pairs of caustic lines is reversed.

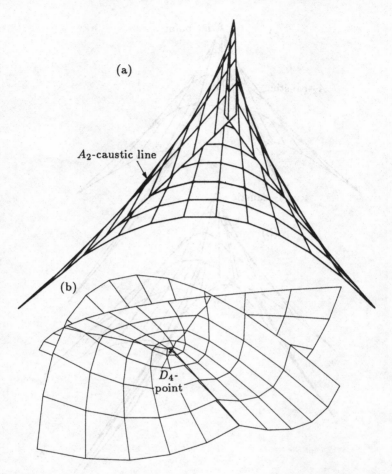

Fig. 4.7.5   The D_4 wavefront when $\epsilon = -1$. (a) The intersection of the wavefront with $t = x^0 = constant < 0$. Shown are two A_2-caustic lines and a self-intersection line which touch at an A_3-caustic point. The A_2-lines have finite length which is $O(t^2)$ as $t \to 0$. There are other self-intersection lines not shown here. (b)The intersection of the wavefront with $t = 0$. The two caustic lines of (a) have collapsed to the D_4-point. Two more self-intersection lines have moved in, so that three meet at the D_4-point. For $t > 0$ the picture becomes the mirror image of (a).

are timelike separated from S and so lie "inside" the spacetime. It should be recalled that every other A_2-caustic in Minkowski spacetime is locally diffeomorphic to this one. Thus we have treated the general case.

An identical calculation can be done for the A_3-case finding

$$ds^2 = -12\epsilon\alpha^2 s_o^4(v_o + 12\epsilon s_o^2 c_o^2), \qquad (4.7.19)$$

and similar results apply. The self-intersection line for this example is given by $x^1 = 0$ or

$$v = -4\epsilon s^2. \qquad (4.7.20)$$

Thus as we move along the generators from $v = 0$ towards the caustic, the self-intersection line is encountered first. One therefore needs to be careful in interpreting fig. 4.7.2. Points on neighbouring generators which are close to the one through the A_3-point are spacelike separated until the caustic is encountered. In particular there is no change in separation as the self-intersection line is crossed. However one must be careful when giving data for the characteristic initial value problem to ensure that they are given consistently at the self-intersection line. Again up to a local diffeomorphism we have treated the general case in Minkowski spacetime. Similar considerations hold for the higher order caustics.

In all of the examples discussed stable caustics occur when the connection coefficients develop simple poles. This was a consequence of the nonlinear equations (4.7.1). Consider next a curved vacuum spacetime, so that we have to replace the second equation (4.7.1) by

$$D\sigma = 2\rho\sigma + \Psi_0,$$

where $\Psi_0 = C_{abcd}l^a m^b l^c m^d$. However this does not change the discussion! In each of the examples discussed we had a spacelike vector e^a which was orthogonal to l^a whose norm tended to zero as the caustic was approached. For example in the A_3-case (4.7.11), $e^a e_a = -(v + 12\epsilon s^2 c^2)^2$. This means that the direction of e^a is approaching that of l^a or equivalently m^a acquires a multiple of l^a

with a factor which tends to infinity as the caustic is approached. However because of the symmetries of C_{abcd} it is manifest that Ψ_0 will remain finite at regular spacetime points and (4.7.5) furnishes the asymptotic forms of ρ and σ at the caustic. There will in general be other, singular Weyl scalars but they do not contain information intrinsic to the caustic.

Appendix A

DIRAC SPINORS

A **Dirac spinor** can be considered as an element ψ^a of a 4-dimensional vector space D isomorphic to $S \oplus \overline{S}^*$. (See section 2.2 for a definition of the spaces S, $\overline{S}^*$.) If κ^A and μ^A are in S we can write

$$\psi^a = \begin{pmatrix} \kappa^A \\ \bar{\mu}_{A'} \end{pmatrix}. \tag{A1}$$

The dual space D^* is obviously isomorphic to $S^* \oplus \overline{S}$, and there are two natural maps : $D \to D^*$. The first is the **Dirac adjoint**:

$$\psi^a \mapsto \bar{\psi}_a = (\mu_A, \quad \bar{\kappa}^{A'}).$$

while the other possibility is **Majorana conjugation**

$$\psi^a \mapsto (\psi_M)_a = (\kappa_A, \quad \bar{\mu}^{A'}).$$

The **Dirac inner product** is

$$\bar{\psi}\psi = 2\Re(\mu_A \kappa^A).$$

A **Majorana spinor** is one for which $\psi_M = \bar{\psi}$.

Because of the structure of D we can decompose Dirac spinors into their 'right' and 'left' parts as follows. Let I_n be the $n \times n$ identity matrix and define

$$\gamma_5 = \begin{pmatrix} I_2 & 0 \\ 0 & -I_2 \end{pmatrix}.$$

For an arbitrary spinor ψ^a the **right** and **left** parts are defined by

$$\psi_R = \tfrac{1}{2}(I_4 + \gamma_5)\psi = \begin{pmatrix} \kappa^A \\ 0 \end{pmatrix}, \qquad \psi_L = \tfrac{1}{2}(I_4 - \gamma_5)\psi = \begin{pmatrix} 0 \\ \bar{\mu}_{A'} \end{pmatrix}.$$

Clearly $\psi = \psi_R + \psi_L$ and the space of right (left) Dirac spinors is isomorphic to S ($\overline{S}^*$).

Now suppose a specific spin basis (o, ι) has been chosen for S, and the associated NP tetrad $(l, n, m, \overline{m})$ and orthonormal tetrad $(e_{\hat{a}})$ defined by equation (2.3.3) have been constructed. Let $\sigma_{\hat{a}}$ be the Pauli spin matrices defined in section 2.1, with $\sigma_{\hat{0}} = I_2$, the identity matrix. As in section 2.3 we define the **Infeld-van der Waerden symbols** via

$$\sigma_{\hat{a}}{}^{AA'} = 2^{-1/2}\sigma_{\hat{a}}. \qquad (A2)$$

Note that index shuffling e.g., $AA' \to A'A$ is not permitted. Further the symmetrization operations indicated below are to apply to tetrad indices only. Lowering and raising of indices is done in the usual way. It is straightforward, but tedious, to verify the following identities

$$\sigma_{(\hat{a}AA'}\sigma_{\hat{b})}{}^{BA'} = \tfrac{1}{2}g_{\hat{a}\hat{b}}\epsilon_A{}^B, \qquad (A3)$$

$$\sigma^{\hat{a}BB'}\sigma_{\hat{a}AA'} = \epsilon_A{}^B\epsilon_{A'}{}^{B'}. \qquad (A4)$$

More generally we might define the van der Waerden symbols axiomatically via equations (A3, A4), and then equation (A2) demonstrates existence.

We may now construct the Dirac γ-operators, where for pedagogic reasons we retain the caret on the 'tensorial' index. Using the ψ given by equation (A1) we define

$$\gamma^{\hat{a}}\psi = \sqrt{2}\begin{pmatrix} \sigma^{\hat{a}AA'}\bar{\mu}_{A'} \\ \sigma^{\hat{a}}{}_{AA'}\kappa^A \end{pmatrix}. \qquad (A5)$$

Written as a 4-matrix $\gamma^{\hat{a}}$ has the form

$$\gamma^{\hat{a}} = \begin{pmatrix} 0 & \sqrt{2}\sigma^{\hat{a}AA'} \\ \sqrt{2}\,[\sigma^{\hat{a}}{}_{AA'}]^T & 0 \end{pmatrix},$$

where T denotes a matrix transpose. Using (A5) and the relations (A2, 3) it is now straightforward to verify

$$\gamma^{(\hat{a}}\gamma^{\hat{b})}\psi = g^{\hat{a}\hat{b}}\psi, \qquad \gamma_{\hat{5}}\psi = i\gamma^{\hat{0}}\gamma^{\hat{1}}\gamma^{\hat{2}}\gamma^{\hat{3}}\psi,$$

from which we obtain the standard relations

$$\gamma^{(\hat{a}}\gamma^{\hat{b})} = g^{\hat{a}\hat{b}}, \qquad \gamma_{\hat{5}} = i\gamma^0\gamma^{\hat{1}}\gamma^{\hat{2}}\gamma^{\hat{3}}. \qquad (A6)$$

The explicit representation (A2) implies

$$\gamma^{\hat{0}} = \begin{pmatrix} 0 & I_2 \\ I_2 & 0 \end{pmatrix}, \qquad \gamma^{\hat{m}} = \begin{pmatrix} 0 & -\sigma_{\hat{m}} \\ \sigma_{\hat{m}} & 0 \end{pmatrix}, \qquad (A7)$$

where $m = 1, 2, 3$. The relations (A6) can also be derived from this explicit form of the γ-matrices.

Appendix B

THE NEWMAN-PENROSE FORMALISM

The connection coefficients

The Newman-Penrose (NP) scalars			
∇	$o^A \nabla o_A$	$o^A \nabla \iota_A = \iota^A \nabla o_A$	$\iota^A \nabla \iota_A$
	$m^a \nabla l_a$	$\frac{1}{2}(n^a \nabla l_a - \bar{m}^a \nabla m_a)$	$-\bar{m}^a \nabla n_a$
D	κ	ϵ	π
Δ	τ	γ	ν
δ	σ	β	μ
$\bar{\delta}$	ρ	α	λ

The commutators acting on a scalar function ψ

$$(\Delta D - D\Delta)\psi = [(\gamma + \bar{\gamma})D + (\epsilon + \bar{\epsilon})\Delta - (\bar{\tau} + \pi)\delta - (\tau + \bar{\pi})\bar{\delta}]\psi$$
$$(\delta D - D\delta)\psi = [(\bar{\alpha} + \beta - \bar{\pi})D + \kappa\Delta - (\bar{\rho} + \epsilon - \bar{\epsilon})\delta - \sigma\bar{\delta}]\psi$$
$$(\delta\Delta - \Delta\delta)\psi = [-\bar{\nu}D + (\tau - \bar{\alpha} - \beta)\Delta + (\mu - \gamma + \bar{\gamma})\delta + \bar{\lambda}\bar{\delta}]\psi$$
$$(\bar{\delta}\delta - \delta\bar{\delta})\psi = [(\bar{\mu} - \mu)D + (\bar{\rho} - \rho)\Delta + (\alpha - \bar{\beta})\delta - (\bar{\alpha} - \beta)\bar{\delta}]\psi$$

The source-free Maxwell Equations

$$D\phi_1 - \bar{\delta}\phi_0 = (\pi - 2\alpha)\phi_0 + 2\rho\phi_1 - \kappa\phi_2 \qquad \text{(Ma)}$$
$$D\phi_2 - \bar{\delta}\phi_1 = -\lambda\phi_0 + 2\pi\phi_1 + (\rho - 2\epsilon)\phi_2 \qquad \text{(Mb)}$$
$$\Delta\phi_0 - \delta\phi_1 = (2\gamma - \mu)\phi_0 - 2\tau\phi_1 + \sigma\phi_2 \qquad \text{(Mc)}$$
$$\Delta\phi_1 - \delta\phi_2 = \nu\phi_0 - 2\mu\phi_1 + (2\beta - \tau)\phi_2 \qquad \text{(Md)}$$

The Newman-Penrose "field" equations

$$D\rho - \bar{\delta}\kappa = (\rho^2 + \sigma\bar{\sigma}) + (\epsilon + \bar{\epsilon})\rho - \bar{\kappa}\tau - \kappa(3\alpha + \bar{\beta} - \pi) + \Phi_{00} \quad \text{(a)}$$

$$D\sigma - \delta\kappa = (\rho + \bar{\rho} + 3\epsilon - \bar{\epsilon})\sigma - (\tau - \bar{\pi} + \bar{\alpha} + 3\beta)\kappa + \Psi_0 \quad \text{(b)}$$

$$D\tau - \Delta\kappa = (\tau + \bar{\pi})\rho + (\bar{\tau} + \pi)\sigma + (\epsilon - \bar{\epsilon})\tau - (3\gamma + \bar{\gamma})\kappa + \Psi_1 + \Phi_{01} \quad \text{(c)}$$

$$D\alpha - \bar{\delta}\epsilon = (\rho + \bar{\epsilon} - 2\epsilon)\alpha + \beta\bar{\sigma} - \bar{\beta}\epsilon - \kappa\lambda - \bar{\kappa}\gamma + (\epsilon + \rho)\pi + \Phi_{10} \quad \text{(d)}$$

$$D\beta - \delta\epsilon = (\alpha + \pi)\sigma + (\bar{\rho} - \bar{\epsilon})\beta - (\mu + \gamma)\kappa - (\bar{\alpha} - \bar{\pi})\epsilon + \Psi_1 \quad \text{(e)}$$

$$D\gamma - \Delta\epsilon = (\tau + \bar{\pi})\alpha + (\bar{\tau} + \pi)\beta - (\epsilon + \bar{\epsilon})\gamma - (\gamma + \bar{\gamma})\epsilon + \tau\pi - \nu\kappa$$
$$+ \Psi_2 - \Lambda + \Phi_{11} \quad \text{(f)}$$

$$D\lambda - \bar{\delta}\pi = (\rho - 3\epsilon + \bar{\epsilon})\lambda + \bar{\sigma}\mu + (\pi + \alpha - \bar{\beta})\pi - \nu\bar{\kappa} + \Phi_{20} \quad \text{(g)}$$

$$D\mu - \delta\pi = (\bar{\rho} - \epsilon - \bar{\epsilon})\mu + \sigma\lambda + (\bar{\pi} - \bar{\alpha} + \beta)\pi - \nu\kappa + \Psi_2 + 2\Lambda \quad \text{(h)}$$

$$D\nu - \Delta\pi = (\pi + \bar{\tau})\mu + (\bar{\pi} + \tau)\lambda + (\gamma - \bar{\gamma})\pi - (3\epsilon + \bar{\epsilon})\nu + \Psi_3 + \Phi_{21} \quad \text{(i)}$$

$$\Delta\lambda - \bar{\delta}\nu = -(\mu + \bar{\mu} + 3\gamma - \bar{\gamma})\lambda + (3\alpha + \bar{\beta} + \pi - \bar{\tau})\nu - \Psi_4 \quad \text{(j)}$$

$$\delta\rho - \bar{\delta}\sigma = (\bar{\alpha} + \beta)\rho - (3\alpha - \bar{\beta})\sigma + (\rho - \bar{\rho})\tau + (\mu - \bar{\mu})\kappa - \Psi_1 + \Phi_{01} \quad \text{(k)}$$

$$\delta\alpha - \bar{\delta}\beta = \mu\rho - \lambda\sigma + \alpha\bar{\alpha} + \beta\bar{\beta} - 2\alpha\beta + (\rho - \bar{\rho})\gamma + (\mu - \bar{\mu})\epsilon - \Psi_2 + \Lambda + \Phi_{11}$$
$$\text{(l)}$$

$$\delta\lambda - \bar{\delta}\mu = (\rho - \bar{\rho})\nu + (\mu - \bar{\mu})\pi + (\alpha + \bar{\beta})\mu + (\bar{\alpha} - 3\beta)\lambda - \Psi_3 + \Phi_{21} \quad \text{(m)}$$

$$\Delta\mu - \delta\nu = -(\mu + \gamma + \bar{\gamma})\mu - \lambda\bar{\lambda} + \bar{\nu}\pi + (\bar{\alpha} + 3\beta - \tau)\nu - \Phi_{22} \quad \text{(n)}$$

$$\Delta\beta - \delta\gamma = (\bar{\alpha} + \beta - \tau)\gamma - \mu\tau + \sigma\nu + \epsilon\bar{\nu} + (\gamma - \bar{\gamma} - \mu)\beta - \alpha\bar{\lambda} - \Phi_{12} \quad \text{(o)}$$

$$\Delta\sigma - \delta\tau = -(\mu - 3\gamma + \bar{\gamma})\sigma - \bar{\lambda}\rho - (\tau + \beta - \bar{\alpha})\tau + \kappa\bar{\nu} - \Phi_{02} \quad \text{(p)}$$

$$\Delta\rho - \bar{\delta}\tau = (\gamma + \bar{\gamma} - \bar{\mu})\rho - \sigma\lambda + (\bar{\beta} - \alpha - \bar{\tau})\tau + \nu\kappa - \Psi_2 - 2\Lambda \quad \text{(q)}$$

$$\Delta\alpha - \bar{\delta}\gamma = (\rho + \epsilon)\nu - (\tau + \beta)\lambda + (\bar{\gamma} - \bar{\mu})\alpha + (\bar{\beta} - \bar{\tau})\gamma - \Psi_3 \quad \text{(r)}$$

The Bianchi identities

$$D\Psi_1 - \bar{\delta}\Psi_0 - D\Phi_{01} + \delta\Phi_{00}$$
$$= (\pi - 4\alpha)\Psi_0 + 2(2\rho + \epsilon)\Psi_1 - 3\kappa\Psi_2 - (\bar{\pi} - 2\bar{\alpha} - 2\beta)\Phi_{00}$$
$$-2(\bar{\rho} + \epsilon)\Phi_{01} - 2\sigma\Phi_{10} + 2\kappa\Phi_{11} + \bar{\kappa}\Phi_{02} \tag{Ba}$$

$$\Delta\Psi_0 - \delta\Psi_1 + D\Phi_{02} - \delta\Phi_{01}$$
$$= (4\gamma - \mu)\Psi_0 - 2(2\tau + \beta)\Psi_1 + 3\sigma\Psi_2 - \bar{\lambda}\Phi_{00} + 2(\bar{\pi} - \beta)\Phi_{01}$$
$$+ 2\sigma\Phi_{11} + (\bar{\rho} + 2\epsilon - 2\bar{\epsilon})\Phi_{02} - 2\kappa\Phi_{12} \tag{Bb}$$

$$D\Psi_2 - \bar{\delta}\Psi_1 + \Delta\Phi_{00} - \bar{\delta}\Phi_{01} + 2D\Lambda$$
$$= -\lambda\Psi_0 + 2(\pi - \alpha)\Psi_1 + 3\rho\Psi_2 - 2\kappa\Psi_3 + (2\gamma + 2\bar{\gamma} - \bar{\mu})\Phi_{00}$$
$$- 2(\alpha + \bar{\tau})\Phi_{01} - 2\tau\Phi_{10} + 2\rho\Phi_{11} + \bar{\sigma}\Phi_{02} \tag{Bc}$$

$$\Delta\Psi_1 - \delta\Psi_2 - \Delta\Phi_{01} + \bar{\delta}\Phi_{02} - 2\delta\Lambda$$
$$= \nu\Psi_0 + 2(\gamma - \mu)\Psi_1 - 3\tau\Psi_2 + 2\sigma\Psi_3 - \bar{\nu}\Phi_{00} + 2(\bar{\mu} - \gamma)\Phi_{01}$$
$$+ (2\alpha + \bar{\tau} - 2\bar{\beta})\Phi_{02} + 2\tau\Phi_{11} - 2\rho\Phi_{12} \tag{Bd}$$

$$D\Psi_3 - \bar{\delta}\Psi_2 - D\Phi_{21} + \delta\Phi_{20} - 2\bar{\delta}\Lambda$$
$$= -2\lambda\Psi_1 + 3\pi\Psi_2 + 2(\rho - \epsilon)\Psi_3 - \kappa\Psi_4 + 2\mu\Phi_{10} - 2\pi\Phi_{11}$$
$$- (2\beta + \bar{\pi} - 2\bar{\alpha})\Phi_{20} - 2(\bar{\rho} - \epsilon)\Phi_{21} + \bar{\kappa}\Phi_{22} \tag{Be}$$

$$\Delta\Psi_2 - \delta\Psi_3 + D\Phi_{22} - \delta\Phi_{21} + 2\Delta\Lambda$$
$$= 2\nu\Psi_1 - 3\mu\Psi_2 + 2(\beta - \tau)\Psi_3 + \sigma\Psi_4 - 2\mu\Phi_{11} - \bar{\lambda}\Phi_{20}$$
$$+ 2\pi\Phi_{12} + 2(\beta + \bar{\pi})\Phi_{21} + (\bar{\rho} - 2\epsilon - 2\bar{\epsilon})\Phi_{22} \tag{Bf}$$

$$D\Psi_4 - \bar{\delta}\Psi_3 + \Delta\Phi_{20} - \bar{\delta}\Phi_{21}$$
$$= -3\lambda\Psi_2 + 2(\alpha + 2\pi)\Psi_3 + (\rho - 4\epsilon)\Psi_4 + 2\nu\Phi_{10} - 2\lambda\Phi_{11}$$
$$- (2\gamma - 2\bar{\gamma} + \mu)\Phi_{20} - 2(\bar{\tau} - \alpha)\Phi_{21} + \bar{\sigma}\Phi_{22} \tag{Bg}$$

$$\Delta\Psi_3 - \delta\Psi_4 - \Delta\Phi_{21} + \bar{\delta}\Phi_{22}$$
$$= 3\nu\Psi_2 - 2(\gamma + 2\mu)\Psi_3 + (4\beta - \tau)\Psi_4 - 2\nu\Phi_{11} - \bar{\nu}\Phi_{20}$$
$$+ 2\lambda\Phi_{12} + 2(\gamma + \bar{\mu})\Phi_{21} + (\bar{\tau} - 2\bar{\beta} - 2\alpha)\Phi_{22} \tag{Bh}$$

$$D\Phi_{11} - \delta\Phi_{10} + \Delta\Phi_{00} - \bar{\delta}\Phi_{01} + 3D\Lambda$$
$$= (2\gamma + 2\bar{\gamma} - \mu - \bar{\mu})\Phi_{00} + (\pi - 2\alpha - 2\bar{\tau})\Phi_{01}$$
$$+ (\bar{\pi} - 2\bar{\alpha} - 2\tau)\Phi_{10} + 2(\rho + \bar{\rho})\Phi_{11} + \bar{\sigma}\Phi_{02}$$
$$+ \sigma\Phi_{20} - \bar{\kappa}\Phi_{12} - \kappa\Phi_{21} \tag{Bi}$$

$$D\Phi_{12} - \delta\Phi_{11} + \Delta\Phi_{01} - \bar{\delta}\Phi_{02} + 3\delta\Lambda$$
$$= (2\gamma - \mu - 2\bar{\mu})\Phi_{01} + \bar{\nu}\Phi_{00} - \bar{\lambda}\Phi_{10}$$
$$+ 2(\bar{\pi} - \tau)\Phi_{11} + (\pi + 2\bar{\beta} - 2\alpha - \bar{\tau})\Phi_{02}$$
$$+ (2\rho + \bar{\rho} - 2\bar{\epsilon})\Phi_{12} + \sigma\Phi_{21} - \kappa\Phi_{22} \tag{Bj}$$

$$D\Phi_{22} - \delta\Phi_{21} + \Delta\Phi_{11} - \bar{\delta}\Phi_{12} + 3\Delta\Lambda$$
$$= \nu\Phi_{01} + \bar{\nu}\Phi_{10} - 2(\mu + \bar{\mu})\Phi_{11} - \lambda\Phi_{02}$$
$$- \bar{\lambda}\Phi_{20} + (2\pi - \bar{\tau} + 2\bar{\beta})\Phi_{12} + (2\beta - \tau + 2\bar{\pi})\Phi_{21}$$
$$+ (\rho + \bar{\rho} - 2\epsilon - 2\bar{\epsilon})\Phi_{22} \tag{Bk}$$

Alternative expressions for the connection coefficients

$$Do_A = \epsilon o_A - \kappa\iota_A \qquad\qquad D\iota_A = \pi o_A - \epsilon\iota_A$$
$$\Delta o_A = \gamma o_A - \tau\iota_A \qquad\qquad \Delta\iota_A = \nu o_A - \gamma\iota_A$$
$$\delta o_A = \beta o_A - \sigma\iota_A \qquad\qquad \delta\iota_A = \mu o_A - \beta\iota_A$$
$$\bar{\delta} o_A = \alpha o_A - \rho\iota_A \qquad\qquad \bar{\delta}\iota_A = \lambda o_A - \alpha\iota_A$$

$$Dl = (\epsilon + \bar{\epsilon})l - \bar{\kappa}m - \kappa\overline{m}$$
$$\Delta l = (\gamma + \bar{\gamma})l - \bar{\tau}m - \tau\overline{m}$$
$$\delta l = (\bar{\alpha} + \beta)l - \bar{\rho}m - \sigma\overline{m}$$
$$Dn = -(\epsilon + \bar{\epsilon})n + \pi m + \bar{\pi}\overline{m}$$
$$\Delta n = -(\gamma + \bar{\gamma})n + \nu m + \bar{\nu}\overline{m}$$
$$\delta n = -(\bar{\alpha} + \beta)n + \mu m + \bar{\lambda}\overline{m}$$
$$Dm = \bar{\pi}l - \kappa n + (\epsilon - \bar{\epsilon})m$$
$$\Delta m = \bar{\nu}l - \tau n + (\gamma - \bar{\gamma})m$$
$$\delta m = \bar{\lambda}l - \sigma n + (\beta - \bar{\alpha})m$$
$$\bar{\delta} m = \bar{\mu}l - \rho n + (\alpha - \bar{\beta})m$$

Behaviour under Lorentz transformations

We give here the behaviour of the NP scalars under the Lorentz transformations discussed in chapter II. Under the spin-boost transformation (2.6.16),

$$(\tilde{o}, \tilde{\iota}) = (\lambda o, \lambda^{-1}\iota), \qquad \lambda = a \exp(i\theta),$$

we have

$$\tilde{l} = a^2 l, \quad \tilde{n} = a^{-2} n, \quad \tilde{m} = e^{2i\theta} m.$$

The NP scalars transform according to:

$$
\begin{aligned}
&\tilde{\kappa} = a^4 e^{2i\theta}\kappa, \quad &&\tilde{\pi} = e^{-2i\theta}\pi, \quad &&\tilde{\epsilon} = a^2(\epsilon + D(\ln a + i\theta)),\\
&\tilde{\tau} = e^{2i\theta}\tau, \quad &&\tilde{\nu} = a^{-4}e^{-2i\theta}\nu, \quad &&\tilde{\gamma} = a^{-2}(\gamma + \Delta(\ln a + i\theta)),\\
&\tilde{\sigma} = a^2 e^{4i\theta}\sigma, \quad &&\tilde{\mu} = a^{-2}\mu, \quad &&\tilde{\beta} = e^{2i\theta}(\beta + \delta(\ln a + i\theta)),\\
&\tilde{\rho} = a^2 \rho, \quad &&\tilde{\lambda} = a^{-2}e^{-4i\theta}\lambda, \quad &&\tilde{\alpha} = e^{-2i\theta}(\alpha + \bar{\delta}(\ln a + i\theta)).
\end{aligned}
$$

$$
\begin{aligned}
&\tilde{\Phi}_{00} = a^4 \Phi_{00}, \quad &&\tilde{\Phi}_{01} = a^2 e^{2i\theta}\Phi_{01}, \quad &&\tilde{\Phi}_{02} = e^{4i\theta}\Phi_{02},\\
&\tilde{\Phi}_{11} = \Phi_{11}, \quad &&\tilde{\Phi}_{12} = a^{-2}e^{2i\theta}\Phi_{12}, \quad &&\tilde{\Phi}_{22} = a^{-4}\Phi_{22}.
\end{aligned}
$$

$$
\begin{aligned}
&\tilde{\Psi}_0 = a^4 e^{4i\theta}\Psi_0, \quad &&\tilde{\Psi}_1 = a^2 e^{2i\theta}\Psi_1, \quad &&\tilde{\Psi}_2 = \Psi_2,\\
&\tilde{\Psi}_3 = a^{-2}e^{-2i\theta}\Psi_3, \quad &&\tilde{\Psi}_4 = a^{-4}e^{-4i\theta}\Psi_4.
\end{aligned}
$$

Under a null rotation about l,

$$(\hat{o}, \hat{\iota}) = (o, \iota + c o),$$

we have

$$\hat{l} = l, \quad \hat{m} = m + \bar{c}l, \quad \hat{n} = n + cm + \bar{c}\bar{m} + c\bar{c}l.$$

The NP scalars transform as

$$
\begin{aligned}
&\hat{\kappa} = \kappa,\\
&\hat{\epsilon} = \epsilon + c\kappa,\\
&\hat{\sigma} = \sigma + \bar{c}\kappa,\\
&\hat{\rho} = \rho + c\kappa,\\
&\hat{\tau} = \tau + c\sigma + \bar{c}\rho + c\bar{c}\kappa,
\end{aligned}
$$

$$\hat{\alpha} = \alpha c \epsilon + c\rho + c^2 \kappa,$$

$$\hat{\beta} = \beta + c\sigma + \bar{c}\epsilon + c\bar{c}\kappa,$$

$$\hat{\pi} = \pi + 2c\epsilon + c^2 \kappa + Dc,$$

$$\hat{\gamma} = \gamma + \bar{c}\alpha + c(\tau + \beta) + c\bar{c}(\rho + \epsilon) + c^2 \sigma + c^2 \bar{c}\kappa,$$

$$\hat{\lambda} = \lambda + c\pi + 2c\alpha + c^2(\rho + 2\epsilon) + c^3 \kappa + cDc + \bar{\delta}c,$$

$$\hat{\mu} = \mu + 2c\beta + \bar{c}\pi + c^2 \sigma + 2c c\bar{c} + c^2 \bar{c}\kappa + \bar{c}Dc + \delta c,$$

$$\hat{\nu} = \nu + c(2\gamma + \mu) + \bar{c}\lambda + c^2(\tau + 2\beta) + c\bar{c}(\pi + 2\alpha) + c^3 \sigma +$$
$$c^2 \bar{c}(\rho + 2\epsilon) + c^3 \bar{c}\kappa + \Delta c + c\delta c + \bar{c}\bar{\delta}c + c\bar{c}Dc.$$

$$\hat{\Phi}_{00} = \Phi_{00},$$

$$\hat{\Phi}_{01} = \Phi_{01} + \bar{c}\Phi_{00},$$

$$\hat{\Phi}_{02} = \Phi_{02} + 2\bar{c}\Phi_{01} + \bar{c}^2 \Phi_{00},$$

$$\hat{\Phi}_{11} = \Phi_{11} + c\Phi_{01} + \bar{c}\Phi_{10} + c\bar{c}\Phi_{00},$$

$$\hat{\Phi}_{12} = \Phi_{12} + c\Phi_{02} + 2\bar{c}\Phi_{11} + 2c\bar{c}\Phi_{01} + \bar{c}^2 \Phi_{10} + c\bar{c}^2 \Phi_{00},$$

$$\hat{\Phi}_{22} = \Phi_{22} + 2c\Phi_{12} + 2\bar{c}\Phi_{21} + c^2 \Phi_{02} + 4c\bar{c}\Phi_{11} + \bar{c}^2 \Phi_{20} +$$
$$2c^2 \bar{c}\Phi_{01} + 2c\bar{c}^2 \Phi_{10} + c^2 \bar{c}^2 \Phi_{00}.$$

$$\hat{\Psi}_0 = \Psi_0,$$

$$\hat{\Psi}_1 = \Psi_1 + c\Psi_0,$$

$$\hat{\Psi}_2 = \Psi_2 + 2c\Psi_1 + c^2 \Psi_0,$$

$$\hat{\Psi}_3 = \Psi_3 + 3c\Psi_2 + 3c^2 \Psi_1 + c^3 \Psi_0,$$

$$\hat{\Psi}_4 = \Psi_4 + 4c\Psi_3 + 6c^2 \Psi_2 + 4c^3 \Psi_1 + c^4 \Psi_0.$$

REFERENCES

Arnol'd, V.I. (1978). *Mathematical Methods of Classical Mechanics* (Springer, New York).

Arnol'd, V.I., Gusein-Zade, S.M. and Varchenko, A.N. (1985). *Singularities of Differentiable Maps I* (Birkhäuser, Boston).

Buchdahl, H.A. (1958). On the compatibility of relativistic wave equations for particles of higher spin in the presence of a gravitational field, *Nuovo Cim.* **10**, 96–103.

Chandrasekhar, S. (1983). *The mathematical theory of black holes* (Oxford University Press, Oxford).

Cocke, W.J. (1989). Table for constructing the spin coefficients in general relativity, *Phys. Rev.* **D40**, 650–651.

Courant, R and Hilbert, D. (1962). *Methods of mathematical physics II* (Interscience, New York).

Duff, G.F.D. (1958). Mixed problems for linear systems of first order equations, *Canadian J. Math.* **10**, 127–160.

Ehlers, J. and Kundt, W. (1962). Exact solutions of the gravitational field equations, in *Gravitation: an introduction to current research*, ed. L. Witten (Wiley, New York).

Fischer, A.E. and Marsden, J.E. (1972). The Einstein evolution equations as a first-order quasi-linear symmetric hyperbolic system I, *Commun. Math. Phys.* **28**, 1–38.

Friedrich, H. (1979). Eine Untersuchung der Einsteinschen Vakuumfeldgleichungen in der Umgebung regulärer und singulärer Nullhyperflächen, Hamburg University Ph.D. thesis.

Friedrich, H. (1981a). On the regular and the asymptotic characteristic initial value problem for Einstein's vacuum field equations, *Proc. Roy. Soc. Lond.* **A375**, 169–184.

Friedrich, H. (1981b). The asymptotic characteristic initial value problem for Einstein's vacuum field equations as an initial value problem for a first order quasilinear symmetric hyperbolic system, *Proc. Roy. Soc. Lond.* **A378**, 401–421.

Friedrich, H. (1982). On the existence of analytic null asymptotically flat solutions of Einstein's vacuum field equations, *Proc. Roy. Soc. Lond.* **A381**, 361–371.

Friedrich, H. (1983). On the hyperbolicity of Einstein's and other gauge field equations, *Commun. Math. Phys.* **100**, 525–543.

Friedrich, H. (1984). On some (con-)formal properties of Einstein's field equations and their consequences, in *Asymptotic behaviour of mass and spacetime geometry*, ed. F.J. Flaherty (Springer, Berlin).

Friedrich, H. and Stewart, J.M. (1983). Characteristic initial data and wave-front singularities in general relativity, *Proc. Roy. Soc. Lond.* **A385,** 345-371.

Geroch, R. (1968). Spinor structure of space-times in general relativity I, *J. Math. Phys.* **9,** 1739-1744.

Goldberg, J.N., Macfarlane, A.J., Newman, E.T., Rohrlich, F. and Sudarshan, E.C.G. (1967). Spin-s spherical harmonics and $\eth$, *J. Math. Phys.* **8,** 2155-2161.

Hawking, S.W. and Ellis, G.F.R. (1973). *The large scale structure of space-time* (Cambridge University Press, Cambridge).

Hawking, S.W. and Israel, W. (1979). *General relativity* (Cambridge University Press, Cambridge).

Hawking, S.W. and Israel, W. (1987). *Three hundred years of gravitation* (Cambridge University Press, Cambridge).

Held, A. (1980). *General relativity and gravitation, volumes 1,2* (Plenum, New York).

Hörmander, L. (1976). *Linear Partial Differential Operators* (Springer, Berlin).

Isaacson, R.A., Welling, J.S. and Winicour, J. (1983). Null cone computation of gravitational radiation, *J. Math. Phys.* **24,** 1824-1834.

Kato, T. (1975). The Cauchy problem for quasi-linear symmetric hyperbolic systems, *Arch. Ration. Mech. Anal.* **58,** 181-205.

Komar, A.B. (1959). Covariant conservation laws in general relativity, *Phys. Rev.* **113,** 934-936.

Kramer, D., Stephani, H., MacCallum, M. and Herlt, E. (1980). *Exact solutions of Einstein's field equations* (Cambridge University Press, Cambridge).

Ludvigsen, M. and Vickers, J.A.G. (1982). A simple proof of the positivity of the Bondi mass, *J. Phys.* **A14,** L67-70.

Misner, C.W., Thorne, K.S. and Wheeler, J.A. (1973). *Gravitation* (Freeman, San Francisco).

Naimark, M.A. (1964). *Linear Representations of the Lorentz Group* (Pergamon, Oxford).

Newman, E.T. and Penrose, R. (1962). An approach to gravitational radiation by a method of spin coefficients, *J. Math. Phys.* **3,** 566-578.

Newman, E.T. and Unti, T.W.J. (1962). Behaviour of asymptotically flat empty space, *J. Math. Phys.* **3,** 891-901.

Penrose, R. (1965). Zero rest-mass fields including gravitation: asymptotic behaviour, *Proc. Roy. Soc. Lond.* **A284,** 159-203.

Penrose, R. and Rindler, W. (1984). *Spinors and space-time 1: Two- spinor calculus and relativistic fields* (Cambridge University Press, Cambridge).

Penrose, R. and Rindler, W. (1986). *Spinors and space-time 2: Spinor & twistor methods in space-time geometry* (Cambridge University Press, Cambridge).

Pirani, F.A.E. (1965). Introduction to gravitational radiation theory, in *Lectures on General Relativity,* ed. Trautman, A., Pirani, F.A.E. and Bondi, H. (Prentice-Hall, Englewood Cliffs).

Rendall, A.D. (1990). Reduction of the characteristic initial value problem to the Cauchy problem and its application to the Einstein equations, *Proc. Roy. Soc. Lond.* **A427,** 221-239.

Sachs, R.K. (1962). On the characteristic initial value problem in gravitation theory, *J. Math. Phys.* **3**, 908–914.

Schouten, J.A. (1954). *Ricci Calculus* (Springer, Berlin).

Stewart, J.M. and Friedrich, H. (1982). Numerical Relativity I: The characteristic initial value problem, *Proc. Roy. Soc. Lond.* **A384**, 427–454.

INDEX